The Benjamin Cummings Team

Selections Editors:

Gerard J. Tortora
Bergen Community College

Berdell R. Funke
North Dakota State University

Christine L. Case
Skyline College

Author, end-of-article material:
Christine L. Case
Skyline College

Executive Editor:
Leslie Berriman
Benjamin Cummings

ON THE COVER

top: Courtesy of Amadeo Bachar
bottom left: Science Source/Photo Researchers, Inc.
bottom center: Tami Tolpa
bottom right: E. Gray *SPL/Photo Researchers, Inc.*

SCIEN
AMEI

Current Issues in Microbiology, Volume 2, is published by
Scientific American, Inc. with project management by:

DIRECTOR, ANCILLARY PRODUCTS: Diane McGarvey
CUSTOM PUBLISHING MANAGER: Marc Richards
CUSTOM PUBLISHING EDITOR: Lisa Pallatroni
DESIGNER, FRONT MATTER: Silvia De Santis

The contents of this issue are adaptations of material
previously published in SCIENTIFIC AMERICAN.

EDITOR IN CHIEF: John Rennie
EXECUTIVE EDITOR: Mariette DiChristina
MANAGING EDITOR: Ricki L. Rusting
CHIEF NEWS EDITOR: Philip M. Yam
SENIOR WRITER: Gary Stix
SENIOR EDITOR: Michelle Press
EDITORS: Mark Alpert, Steven Ashley, Graham P. Collins,
Mark Fischetti, Steve Mirsky, George Musser, Christine Soares

ART DIRECTOR: Edward Bell
ASSOCIATE ART DIRECTOR: Mark Clemens
ASSISTANT ART DIRECTOR: Johnny Johnson
PHOTOGRAPHY EDITOR: Emily Harrison
PRODUCTION EDITOR: Richard Hunt

COPY DIRECTOR: Maria-Christina Keller
COPY CHIEF: Daniel C. Schlenoff
COPY AND RESEARCH: Michael Battaglia, Smitha Alampur,
Michelle Wright, John Matson, Aaron Shattuck

EDITORIAL ADMINISTRATOR: Jacob Lasky
SENIOR SECRETARY: Maya Harty

ASSOCIATE PUBLISHER, PRODUCTION: William Sherman
PREPRESS AND QUALITY MANAGER: Silvia De Santis
PRODUCTION MANAGER: Christina Hippeli

ASSOCIATE PUBLISHER, CIRCULATION: Simon Aronin

ASSOCIATE PUBLISHER, STRATEGIC PLANNING: Laura Salant

GENERAL MANAGER: Michael Florek
BUSINESS MANAGER: Marie Maher

CHAIRMAN: Brian Napack

VICE PRESIDENT AND PUBLISHER: Bruce Brandfon

VICE PRESIDENT: Frances Newburg

Current Issues in Microbiology, Volume 2, published by Scientific American, Inc., 415
Madison Avenue, New York, NY 10017-1111. Copyright © 2007 by Scientific American,
Inc. All rights reserved. No part of this issue may be reproduced by any mechanical,
photographic or electronic process, or in the form of a phonographic recording, nor may
it be stored in a retrieval system, transmitted or otherwise copied for public or private
use without written permission of the publisher.

Subscription inquiries for SCIENTIFIC AMERICAN magazine:
U.S. and Canada (800) 333-1199; other (515) 247-7631, or www.sciam.com.

To learn more about Scientific American's
Custom Publishing Program, contact Marc Richards at 212-451-8859 or
mrichards@sciam.com.

INITIAL ENCOUNTER with germs, or pathogens, sets off the "innate" arm of the immune system, which turns out to be more sophisticated than anyone guessed.

Immunity's Early-Warning System

By Luke A. J. O'Neill

The innate immune response constitutes the first line of defense against invading microbes and plays a role in inflammatory disease. Surprising insights into how this system operates could lead to new therapies for a host of infectious and immune-related disorders

A woman is riding an elevator when her fellow passengers start to sneeze. As she wonders what sort of sickness the other riders might be spreading, her immune system swings into action. If the bug being dispersed by the contagious sneezers is one the woman has met before, a battalion of trained immune cells—the foot soldiers of the so-called adaptive immune system—will remember the specific invader and clear it within hours. She might never realize she had been infected.

But if the virus or bacterium is one that our hapless rider has never wrestled, a different sort of immune response comes to the rescue. This "innate" immune system recognizes generic classes of molecules produced by a variety of disease-causing agents, or pathogens. When such foreign molecules are detected, the innate system triggers an inflammatory response, in which certain cells of the immune system attempt to wall off the invader and halt its spread. The activity of these cells—and of the chemicals they secrete—precipitates the redness and swelling at sites of injury and accounts for the fever, body aches and other flulike symptoms that accompany many infections.

The inflammatory assault, we now know, is initiated by Toll-like receptors (TLRs): an ancient family of proteins that mediate innate immunity in organisms from horseshoe crabs to humans. If TLRs fail, the entire immune system crashes, leaving the body wide open to infection. If they work too hard, however, they can induce disorders marked by chronic, harmful inflammation, such as arthritis, lupus and even cardiovascular disease.

Discovery of TLRs has generated an excitement among immunologists akin to that seen when Christopher Columbus returned from the New World. Scores of researchers are now setting sail to this new land, where they hope to find explanations for many still mysterious aspects of immunity, infection and disorders involving abnormal defensive activity. Study of these receptors, and of the molecular events that unfold after they encounter a pathogen, is already beginning to uncover targets for pharmaceuticals that may enhance the body's protective activity, bolster vaccines, and treat a range of devastating and potentially deadly disorders.

Cinderella Immunity

UNTIL ABOUT FIVE YEARS AGO, when it came to the immune system, the adaptive division was the star of the show. Textbooks were filled with details about B cells making antibodies that latch onto specific proteins, or antigens, on the surface of an invading pathogen and about T cells that sport receptors able to recognize fragments of proteins from pathogens. The response is called adaptive because over the course of an infection, it adjusts to optimally handle the particular microorganism responsible for the disease.

Adaptive immunity also grabbed the spotlight because it endows the immune system with memory. Once an infection has been eliminated, the specially trained B and T cells stick around, priming the body to ward off subsequent attacks. This ability to remember past infections allows vaccines to protect us from diseases caused by viruses or bacteria. Vac-

The mammalian immune system has two overarching divisions. The innate part (*left side*) acts near entry points into the body and is always at the ready. If it fails to contain a pathogen, the adaptive division (*right side*) kicks in, mounting a later but highly targeted attack against the specific invader.

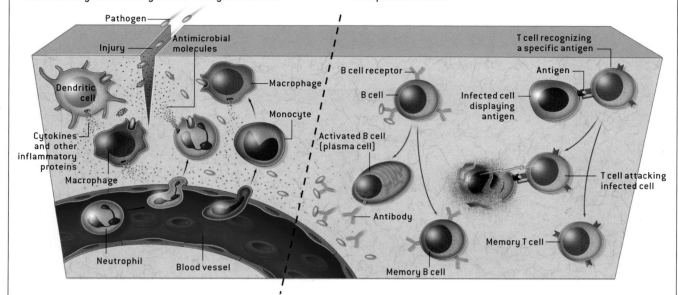

INNATE IMMUNE SYSTEM

This system includes, among other components, antimicrobial molecules and various phagocytes (cells that ingest and destroy pathogens). These cells, such as dendritic cells and macrophages, also activate an inflammatory response, secreting proteins called cytokines that trigger an influx of defensive cells from the blood. Among the recruits are more phagocytes—notably monocytes (which can mature into macrophages) and neutrophils.

ADAPTIVE IMMUNE SYSTEM

This system "stars" B cells and T cells. Activated B cells secrete antibody molecules that bind to antigens—specific components unique to a given invader—and destroy the invader directly or mark it for attack by others. T cells recognize antigens displayed on cells. Some T cells help to activate B cells and other T cells (*not shown*); other T cells directly attack infected cells. T and B cells spawn "memory" cells that promptly eliminate invaders encountered before.

cines expose the body to a disabled form of a pathogen (or harmless pieces of it), but the immune system reacts as it would to a true assault, generating protective memory cells in the process. Thanks to T and B cells, once an organism has encountered a microbe and survived, it becomes exempt from being overtaken by the same bug again.

The innate immune system seemed rather drab in comparison. Its compo-

nents—including antibacterial enzymes in saliva and an interlocking set of proteins (known collectively as the complement) that kill bacteria in the bloodstream—were felt to be less sophisticated than targeted antibodies and killer T cells. What is more, the innate immune system does not tailor its response in the same way that the adaptive system does.

In dismissing the innate immune response as dull and uninteresting, how-

ever, immunologists were tiptoeing around a dirty little secret: the adaptive system does not work in the absence of the allegedly more crude innate response. The innate system produces certain signaling proteins called cytokines that not only induce inflammation but also activate the B and T cells that are needed for the adaptive response. The posh sister, it turns out, needs her less respected sibling to make her shine.

By the late 1990s immunologists knew a tremendous amount about how the adaptive immune system operates. But they had less of a handle on innate immunity. In particular, researchers did not understand how microbes activate the innate response—or exactly how this stimulation helps to drive the adaptive response of T and B cells. Soon after, though, they would learn that much of the answer lay with the TLRs, which are produced by various immune system cells. But the path scientists traveled to get to these proteins was a circuitous one,

Overview/*Innate Immunity*

- Innate immunity serves as a rapid response system for detecting and clearing infections by any infectious agent. The response is mediated by a family of molecules called Toll-like receptors (TLRs), made by many defensive cells.
- When TLRs detect an invader, they trigger the production of an array of signaling proteins that induce inflammation and direct the body to mount a full-fledged immune response.
- If TLRs are underactive, the immune system fails; if overactive, they can give rise to disorders such as rheumatoid arthritis and even cardiovascular disease. Learning how to manipulate TLRs or the proteins with which they interact could provide new options for treating infectious and inflammatory diseases.

winding through studies of fruit fly development, the search for drugs to treat arthritis, and the dawn of the genomic era.

Weird Protein

THE PATH ACTUALLY HAD its beginnings in the early 1980s, when immunologists started to study the molecular activity of cytokines. These protein messengers are produced by various immune cells, including macrophages and dendritic cells. Macrophages patrol the body's tissues, searching for signs of infection. When they detect a foreign protein, they set off the inflammatory response. In particular, they engulf and destroy the invader bearing that protein and secrete a suite of cytokines, some of which raise an alarm that recruits other cells to the site of infection and puts the immune system in general on full alert. Dendritic cells ingest invading microbes and head off to the lymph nodes, where they present fragments of the pathogen's proteins to armies of T cells and release cytokines—activities that help to switch on the adaptive immune response.

To study the functions of various cytokines, researchers needed a way to induce the molecules' production. They found that the most effective way to get macrophages and dendritic cells to make cytokines in the laboratory was to expose them to bacteria—or more important, to selected components of bacteria. Notably, a molecule called lipopolysaccharide (LPS), made by a large class of bacteria, stimulates a powerful immune response. In humans, exposure to LPS causes fever and can lead to septic shock—a deadly vascular shutdown triggered by an overwhelming, destructive action of immune cells. LPS, it turns out, evokes this inflammatory response by prompting macrophages and dendritic cells to release the cytokines tumor necrosis factor-alpha (TNF-alpha) and interleukin-1 (IL-1).

Indeed, these two cytokines were shown to rule the inflammatory response, prodding immune cells into action. If left unchecked, they can precipitate disorders such as rheumatoid arthritis, an autoimmune condition characterized by excessive inflammation that leads to destruc-

TOLLS IN CHARGE

Toll-like receptors (TLRs), made by many cells of the innate system, have been found to both orchestrate the innate immune response and play a critical role in the adaptive response. TLR4, for example, elicits these defenses when gram-negative bacteria begin to invade. TLR4 detects the incursions by binding to lipopolysaccharide (LPS), a sugar unique to gram-negative bacteria. Having recognized LPS, pairs of TLR4s signal to four molecules inside the cell—MyD88, Mal, Tram and Trif—which, in turn, trigger molecular interactions that ultimately activate a master regulator of inflammation (NF-κB). This regulator then switches on genes that encode immune activators, including cytokines. These cytokines (*far right*) induce inflammation and also help to switch on the T and B cells of the adaptive immune division.

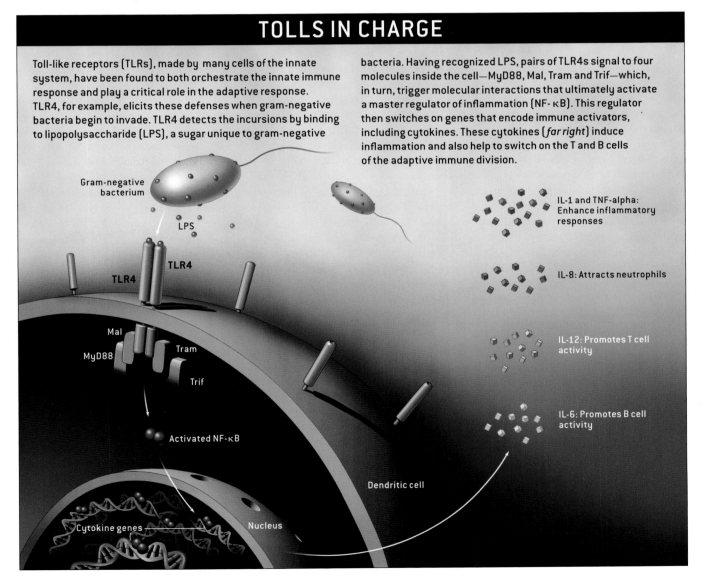

Gram-negative bacterium

LPS

TLR4

TLR4

Mal

MyD88

Tram

Trif

Activated NF-κB

Cytokine genes

Nucleus

Dendritic cell

IL-1 and TNF-alpha: Enhance inflammatory responses

IL-8: Attracts neutrophils

IL-12: Promotes T cell activity

IL-6: Promotes B cell activity

tion of the joints. Investigators therefore surmised that limiting the effects of TNF-alpha and IL-1 might slow the progress of the disease and alleviate the suffering of those with arthritis. To design such a therapy, though, they needed to know more about how these molecules work. And the first step was identifying the proteins with which they interact.

In 1988 John E. Sims and his colleagues at Immunex in Seattle discovered a receptor protein that recognizes IL-1. This receptor resides in the membranes of many different cells in the body, including macrophages and dendritic cells. The part of the receptor that juts out of the cell binds to IL-1, whereas the segment that lies inside the cell relays the message that IL-1 has been detected. Sims examined the inner part of the IL-1 receptor carefully, hoping it would yield some clue as to how the protein transmits its message—revealing, for example, which signaling molecules it activates within cells. But the inner domain of the human IL-1 receptor was unlike anything researchers had seen before, so he was stymied.

Then, in 1991, Nick J. Gay of the University of Cambridge—working on a completely unrelated problem—made a strange discovery. He was looking for proteins that were similar to a fruit-fly protein called Toll. Toll had been identified by Christiane Nusslein-Volhard in Tübingen, Germany, who gave the protein its name because flies that lack Toll look weird (*Toll* being the German word for "weird"). The protein helps the developing *Drosophila* embryo to differentiate its top from its bottom, and flies without Toll look jumbled, as if they have lost their sidedness.

Gay searched the database containing all the gene sequences then known. He

LUKE A. J. O'NEILL received his Ph.D. in pharmacology from the University of London in 1985 for work on the pro-inflammatory cytokine interleukin-1. O'Neill is Science Foundation Ireland Research Professor and head of the department of biochemistry at Trinity College in Dublin. He is founder of Opsona Therapeutics, a drug development company in Dublin.

FRUIT FLY that lacked the protein Toll fell victim to a rampant fungal infection; spores cover the body like a fur coat. (The head is at the bottom right.) This outcome, reported in 1996, was one of the first indications that fruit flies require Toll proteins for protection against disease.

was looking for genes whose sequences closely matched that of Toll and thus might encode Toll-like proteins. And he discovered that part of the Toll protein bears a striking resemblance to the inner part of the human IL-1 receptor, the segment that had mystified Sims.

At first the finding didn't make sense. Why would a protein involved in human inflammation look like a protein that tells fly embryos which end is up? The discovery remained puzzling until 1996, when Jules A. Hoffmann and his collaborators at CNRS in Strasbourg showed that flies use their Toll protein to defend themselves from fungal infection. In *Drosophila*, it seems, Toll multitasks and is involved in both embryonic development and adult immunity.

Worms, Water Fleas and You

THE IL-1 RECEPTOR and the Toll protein are similar only in the segments that are tucked inside the cell; the bits that are exposed to the outside look quite different. This observation led researchers to search for human proteins that resemble Toll in its entirety. After all, evolution usually conserves designs that work well—and if Toll could mediate immunity in flies, perhaps similar proteins were doing the same in humans.

Acting on a tip from Hoffmann, in 1997 Ruslan Medzhitov and the late Charles A. Janeway, Jr., of Yale University discovered the first of these proteins, which they called human Toll. Within six months or so, Fernando Bazan and his

colleagues at DNAX in Palo Alto, Calif., had identified five human Tolls, which they dubbed Toll-like receptors (TLRs). One, TLR4, was the same human Toll described by Medzhitov and Janeway.

At that point, researchers still did not know exactly how TLRs might contribute to human immunity. Janeway had found that stuffing the membranes of dendritic cells with TLR4 prompted the production of cytokines. But he could not say how TLR4 became activated during an infection.

The answer came in late 1998, when Bruce Beutler and his co-workers at the Scripps Institute in La Jolla, Calif., found that mutant mice unable to respond to LPS harbor a defective version of TLR4. Whereas normal mice die of sepsis within an hour of being injected with LPS, these mutant mice survive and behave as if they have not been exposed to the molecule at all; that is, the mutation in the TLR4 gene renders these mice insensitive to LPS.

This discovery made it clear that TLR4 becomes activated when it interacts with LPS. Indeed, its job is to sense LPS. That realization was a major breakthrough in the field of sepsis, because it revealed the molecular mechanism that underlies inflammation and provided a possible new target for treatment of a disorder that sorely needed effective therapies. Within two years, researchers determined that most TLRs—of which 10 are now known in humans—recognize molecules important to the survival of bacteria, viruses, fungi and parasites. TLR2 binds to lipoteichoic acid, a component of the bacterial cell wall. TLR3 recognizes the genetic material of viruses; TLR5 recognizes flagellin, a protein that forms the whiplike tails used by bacteria to swim; and TLR9 recognizes a signature genetic sequence called CpG, which occurs in bacteria and viruses in longer stretches and in a form that is chemically distinct from the CpG sequences in mammalian DNA.

TLRs, it is evident, evolved to recognize and respond to molecules that are fundamental components of pathogens. Eliminating or chemically altering any one of these elements could cripple an infectious agent, which means that the organisms cannot dodge TLRs by mutating until these components are unrecognizable. And because so many of these elements are shared by a variety of microbes, even as few as 10 TLRs can protect us from virtually every known pathogen.

Innate immunity is not unique to humans. In fact, the system is quite ancient. Flies have an innate immune response, as do starfish, water fleas and almost every organism that has been examined thus far. And many use TLRs as a trigger. The nematode worm has one that allows it to sense and swim away from infectious bacteria. And plants are rife with TLRs. Tobacco has one called N protein that is required for fighting tobacco mosaic virus. The weed *Arabidopsis* has more than 200. The first Toll-like protein most likely arose in a single-celled organism that was a common ancestor of plants and animals. Perhaps these molecules even helped to facilitate our evolution. Without an efficient means of defense

against infection, multicellular organisms might never have survived.

Storming the Castle

THE INNATE SYSTEM was once thought to be no more elaborate than the wall of a castle. The real action, researchers believed, occurred once the wall had been breached and the troops inside—the T and B cells—became engaged. We now know that the castle wall is studded with sentries—TLRs—that identify the invader and sound the alarm to mobilize the troops and prepare the array of defenses needed to fully combat the attack. TLRs, in other words, unleash both the innate and adaptive systems.

The emerging picture looks something like this. When a pathogen first enters the body, one or more TLRs, such as those on the surface of patrolling macrophages and dendritic cells, latch onto the foreign molecules—for example, the LPS of gram-negative bacteria. Once en-

gaged, the TLRs prompt the cells to unleash particular suites of cytokines. These protein messengers then recruit additional macrophages, dendritic cells and other immune cells to wall off and nonspecifically attack the marauding microbe. At the same time, cytokines released by all these busy cells can produce the classic symptoms of infection, including fever and flulike feelings.

Macrophages and dendritic cells that have chopped up a pathogen display pieces of it on their surface, along with other molecules indicating that a disease-causing agent is present. This display, combined with the cytokines released in response to TLRs, ultimately activates B and T cells that recognize those specific pieces, causing them—over the course of several days—to proliferate and launch a powerful, highly focused attack on the particular invader. Without the priming effect of TLRs, B and T cells would not become engaged and the body would not

THE JOBS OF TOLL-LIKE RECEPTORS

Each Toll-like receptor can detect some essential component of a broad class of disease-causing agents, and as a group, TLRs can apparently recognize almost every pathogen likely to cause infections. Different combinations appear in different kinds of cells, where the molecules act in pairs. Investigators have identified 10 human TLRs and know many of the molecules they recognize (*below*). The function of TLR10 and the partners of TLR3, 5, 7, 8 and 9 are unknown.

Cell surface

TLR4 TLR4 TLR5

TLR6

TLR2

Binds to LPS from gram-negative bacteria

Binds to flagellin in the tails of motile bacteria

TLR2

TLR1

Binds to lipoteichoic acid (a component of gram-positive bacteria) and to zymosan, produced by fungi

Binds to uniquely bacterial lipopeptides (lipid-protein combinations) and to molecules called GPI-anchored proteins in parasites

Binds to single-stranded viral RNA

Binds to so-called CpG DNA from bacteria or viruses

Binds to single-stranded viral RNA, such as in HIV

Compartment for degrading pathogens

TLR8 TLR9

Binds to double-stranded viral RNA, such as in West Nile virus

TLR7

TLR3

be able to mount a full immune response. Nor could the body retain any memory of previous infections.

Following the initial infection, enough memory T and B cells are left behind so that the body can deal more efficiently with the invader should it return. This army of memory cells can act so quickly that inflammation might not occur at all. Hence, the victim does not feel as ill and might not even notice the infection when it recurs.

Innate and adaptive immunity are thus part of the same system for recognizing and eliminating microbes. The interplay between these two systems is what makes our overall immune system so strong.

Choose Your Weapon

TO FULLY UNDERSTAND how TLRs control immune activity, immunologists need to identify the molecules that relay signals from activated TLRs on the cell surface to the nucleus, switching on genes that encode cytokines and other immune activators. Many investigators are now pursuing this search intensively, but already we have made some fascinating discoveries.

We now know that TLRs, like many receptors that reside on the cell surface, enlist the help of a long line of signaling proteins that carry their message to the nucleus, much as a bucket brigade shuttles water to a fire. All the TLRs, with the exception of TLR3, hand off their signal to an adapter protein called MyD88. Which other proteins participate in the relay varies with the TLR: my laboratory studies Mal, a protein we discovered that helps to carry signals generated by TLR4 and TLR2. TLR4 also requires two other proteins—Tram and Trif—to relay the signal, whereas TLR3 relies on Trif alone. Shizuo Akira of Osaka University in Japan has shown that mice engineered so that they do not produce some of these intermediary signaling proteins do not respond to microbial products, suggesting that TLR-associated proteins could provide novel targets for new anti-inflammatory or antimicrobial agents.

Interaction with different sets of signaling proteins allows TLRs to activate different sets of genes that hone the cell's response to better match the type of pathogen being encountered. For example, TLR3 and TLR7 sense the presence of viruses. They then trigger a string of molecular interactions that induce the production and release of interferon, the major antiviral cytokine. TLR2, which is activated by bacteria, stimulates the release of a blend of cytokines that does not include interferon but is more suited to activating an effective antibacterial response by the body.

The realization that TLRs can detect different microbial products and help to tailor the immune response to thwart the enemy is now overturning long-held assumptions that innate immunity is a static, undiscriminating barrier. It is, in fact, a dynamic system that governs almost every aspect of inflammation and immunity.

From *Legionella* to Lupus

ON RECOGNIZING the central role that TLRs play in initiating immune responses, investigators quickly began to suspect that hobbled or overactive versions of these receptors could contribute to many infectious and immune-related disorders. That hunch proved correct. Defects in innate immunity lead to greater susceptibility to viruses and bacteria. People with an underactive form of TLR4 are five times as likely to have severe bacterial infections over a five-year period than those with a normal TLR4. And people who die from Legionnaire's disease often harbor a mutation in TLR5 that disables the protein, compromising their innate immune response and rendering them unable to fight off the *Legionella* bacterium. On the other hand, an overzealous immune response can be equally destructive. In the U.S. and Europe alone, more than 400,000 people die annually from sepsis, which stems from an overactive immune response led by TLR4.

Other studies are pointing to roles for TLRs in autoimmune diseases such as systemic lupus erythematosus and rheumatoid arthritis. Here TLRs might respond to products from damaged cells, propagating an inappropriate inflammatory response and promoting a misguided reaction by the adaptive immune system. In lupus, for example, TLR9 has been found to react to the body's own DNA.

Innate immunity and the TLRs could also play a part in heart disease. People with a mutation in TLR4 appear to be less prone to developing cardiovascular disease. Shutting down TLR4 could protect the heart because inflammation appears to contribute to the formation of the plaques that clog coronary arteries.

MECHNIKOV'S FLEAS

The discovery of Tolls and Toll-like receptors extends a line of research begun more than 100 years ago, when Russian biologist Ilya Mechnikov essentially discovered innate immunity. In the early 1880s Mechnikov plucked some thorns from a tangerine tree and poked them into a starfish larva. The next morning he saw that the thorns were surrounded by mobile cells, which he surmised were in the process of engulfing bacteria introduced along with the foreign bodies. He then discovered that water fleas (*Daphnia*) exposed to fungal spores mount a similar response. This process of phagocytosis is a cornerstone of innate immunity, and its discovery earned Mechnikov a Nobel Prize in 1908.

MECHNIKOV was a character. Speaking of the era when he worked at the Pasteur Institute, his Nobel Prize biography notes, "It is said of him that at this time he usually wore overshoes in all weathers and carried an umbrella, his pockets being overfull with scientific papers, and that he always wore the same hat, and often, when he was excited, sat on it."

TLRs AS DRUG TARGETS

Agents that activate TLRs and thus enhance immune responses could increase the effectiveness of vaccines or protect against infection. They might even prod the immune system to destroy tumors. In contrast, drugs that block TLR activity might prove useful for dampening inflammatory disorders. Drugs of both types are under study (*below*).

DRUG TYPE	EXAMPLES
TLR4 activator	MPL, an allergy treatment and vaccine adjuvant (immune system activator) from Corixa (Seattle), is in large-scale clinical trials
TLR7 activator	ANA245 (isatoribine), an antiviral agent from Anandys (San Diego), is in early human trials for hepatitis C
TLR7 and TLR8 activator	Imiquimod, a treatment for genital warts, basal cell skin cancer and actinic keratosis from 3M (St. Paul, Minn.), is on the market
TLR9 activator	ProMune, a vaccine adjuvant and treatment for melanoma skin cancer and non-Hodgkin's lymphoma from Coley (Wellesley, Mass.), is in large-scale clinical trials
TLR4 inhibitor	E5564, an antisepsis drug from Eisai (Teaneck, N.J.), is in early human trials
General TLR inhibitor	RDP58, a drug for ulcerative colitis and Crohn's disease from Genzyme (Cambridge, Mass.), is entering large-scale clinical trials
General TLR inhibitor	OPN201, a drug for autoimmune disorders from Opsona Therapeutics (Dublin, Ireland), is being tested in animal models of inflammation

Manipulation of TLR4 might therefore be another approach to preventing or limiting this condition.

Volume Control

MANY OF THE BIG pharmaceutical companies have an interest in using TLRs and their associated signaling proteins as targets for drugs that could treat infections and immune-related disorders. With the spread of antibiotic resistance, the emergence of new and more virulent viruses, and the rising threat of bioterrorism, the need to come up with fresh ways to help our bodies fight infection is becoming more pressing.

Work on TLRs could, for example, guide the development of safer, more effective vaccines. Most vaccines depend on the inclusion of an adjuvant, a substance that kick-starts the inflammatory response, which in turn pumps up the ability of the adaptive system to generate the desired memory cells. The adjuvant used in most vaccines today does not provoke a full adaptive response; instead it favors B cells over T cells. To elicit a stronger response, several companies have set their sights on compounds that

activate TLR9, a receptor that recognizes a broad range of bacteria and viruses and drives a robust immune response.

And TLRs are teaching us how to defend ourselves against biological weapons, such as poxviruses. A potential staple in the bioterrorist arsenal, these viruses can shut down TLRs and thereby avoid detection and elimination. In collaboration with Geoffrey L. Smith of Imperial College London, my lab found that by removing the viral protein that disables TLRs, we could generate a weakened virus that could serve as the basis of a vaccine unlikely to provoke an unintended fatal pox infection.

Armed with an understanding of TLRs and innate immunity, physicians might be able to predict which patients will fare poorly during infection and

treat them more aggressively. If, for instance, patients came to a clinic with a bacterial infection and were found to have a mutant TLR4, the doctor might bombard them with antibiotics or with agents that could somehow bolster their immune response to prevent the infection from doing lasting damage.

Of course, a balance must be struck between stimulating an immune response that is sufficient to clear a microbe and precipitating an inflammatory response that will do more harm than good. Similarly, any medications that aim to relieve inflammation by quelling TLR activity and cytokine release must not, at the same time, undercut the body's defense against infection.

Anti-inflammatory drugs that interfere with TNF-alpha, one of the cytokines produced as a result of TLR4 activation, offer a cautionary tale. TNF-alpha produced during infection and inflammation can accumulate in the joints of patients with rheumatoid arthritis. The anti-inflammatory compounds alleviate the arthritis, but some people taking them wind up with tuberculosis. The infection is probably latent, but reining in the inflammatory response can also dampen the pathogen-specific responses and allow the bacterium to reemerge.

In short, TLRs are like the volume knob on a stereo, balancing adaptive immunity and inflammation. Researchers and pharmaceutical companies are now looking for ways to tweak these controls, so they can curtail inflammation without disabling immunity.

Given that TLRs were unheard of seven years ago, investigators have made enormous progress in understanding the central role these proteins play in the body's first line of defense. Innate immunity, long shrouded in oblivion, has suddenly become the belle of the ball. ☒

MORE TO EXPLORE

Innate Immunity. Ruslan Medzhitov and Charles Janeway in *New England Journal of Medicine*, Vol. 343, No. 5, pages 338–344; August 3, 2000.

Inferences, Questions and Possibilities in Toll-like Receptor Signaling. Bruce Beutler in *Nature*, Vol. 430, pages 257–263; July 8, 2004.

Toll-like Receptor Control of the Adaptive Immune Responses. Akiko Iwasaki and Ruslan Medzhitov in *Nature Immunology*, Vol. 5, No. 10, pages 987–995; October 2004.

TLRs: Professor Mechnikov, Sit on Your Hat. L.A.J. O'Neill in *Trends in Immunology*, Vol. 25, No. 12, pages 687–693; December 2004.

Questions for Review

"Immunity's Early-Warning System"

by Luke A.J. O'Neill

TESTING YOUR COMPREHENSION

1. Innate immunity
   a. involves antibodies.
   b. involves T cells.
   c. is specific for a particular pathogen.
   d. involves macrophages and dendritic cells.

2. Toll-like receptor 4 (TLR4) binds to LPS. Therefore TLR4 is activated by
   a. gram-positive bacteria.
   b. all bacteria.
   c. gram-negative bacteria.
   d. viruses.

3. TLR2 binds to lipoteichoic acid. Therefore TLR2 is activated by
   a. gram-positive bacteria.
   b. all bacteria.
   c. gram-negative bacteria.
   d. viruses.

4. TLR3 and TLR7 induce interferon production. You can conclude that TLR3 and TLR7 are activated by
   a. gram-positive bacteria.
   b. all bacteria.
   c. gram-negative bacteria.
   d. viruses.

5. An adjuvant
   a. improves the immune response.
   b. is activated by TLR5.
   c. produces TLR5.
   d. is activated by LPS.

6. Lupus is an autoimmune disease that results from TLR9 reacting with
   a. human DNA.
   b. bacterial DNA.
   c. viral DNA.
   d. viral RNA.

7. The following occur during activation of the immune system. Which is the second step?
   a. Antigen displayed on a macrophage
   b. Cytokine
   c. Fever
   d. Bacteria ingested by a macrophage

8. People who die from legionellosis have a mutation in
   a. TLR2.
   b. TLR3.
   c. TLR4.
   d. TLR5.

9. People with underactive TLR4 are five times more likely to have
   a. viral infections.
   b. bacterial infections.
   c. autoimmune disease.
   d. arthritis.

10. The trial drug ROP58 is a TLR inhibitor. This drug may be useful to treat
    a. bacterial infections.
    b. Crohn's inflammation.
    c. legionellosis.
    d. viral infections.

MICROBIOLOGY IN SOCIETY

1. Recent evidence suggests that TLRs can be activated by molecules on mammalian cells and therefore contribute to organ transplant rejection. What would be the consequences of shutting down a patient's TLRs to improve transplant success?

2. Infections, including tuberculosis, are possible side effects of nonsteroidal anti-inflammatory drugs (NSAIDs) used to treat arthritis pain. Should NSAIDs be allowed if communicable diseases could occur?

3. The author states "more than 400,000 people die annually from sepsis, which stems from an overactive immune response." If immunity protects you, why do these deaths occur?

THINKING ABOUT MICROBIOLOGY

1. What accounts for the presence of TLRs in every plant and animal studied?

2. If antibodies and T cells destroy pathogens, why is innate immunity important?

3. Dendritic cells of the liver have fewer TLR4 molecules than other organs such as the spleen. Consider the position of the liver and function of TLR4 to explain the possible value of fewer TLR4 molecules.

WRITING ABOUT MICROBIOLOGY

1. TLRs are the early-warning signal for the immune system. How did toll-like receptors get their unusual name?

2. Agonists are drugs that bind to a cell-surface receptor. Imiquimod is a TLR7 agonist in a topical cream to treat genital warts. Explain the mechanism of action of this drug.

3. The author states "Shutting down TLR4 could protect the heart because inflammation appears to contribute to the formation of the plaques that clog coronary arteries." What considerations should a physician and patient discuss to evaluate possible benefits against adverse reactions?

Answers can be found on The Microbiology Place website. Go to www.microbiologyplace.com, click on the cover of your textbook, and type in your login name and password (using the access code found in the front pages of your textbook). Then, click on Current Issues Magazine Answers on the left navigation bar.

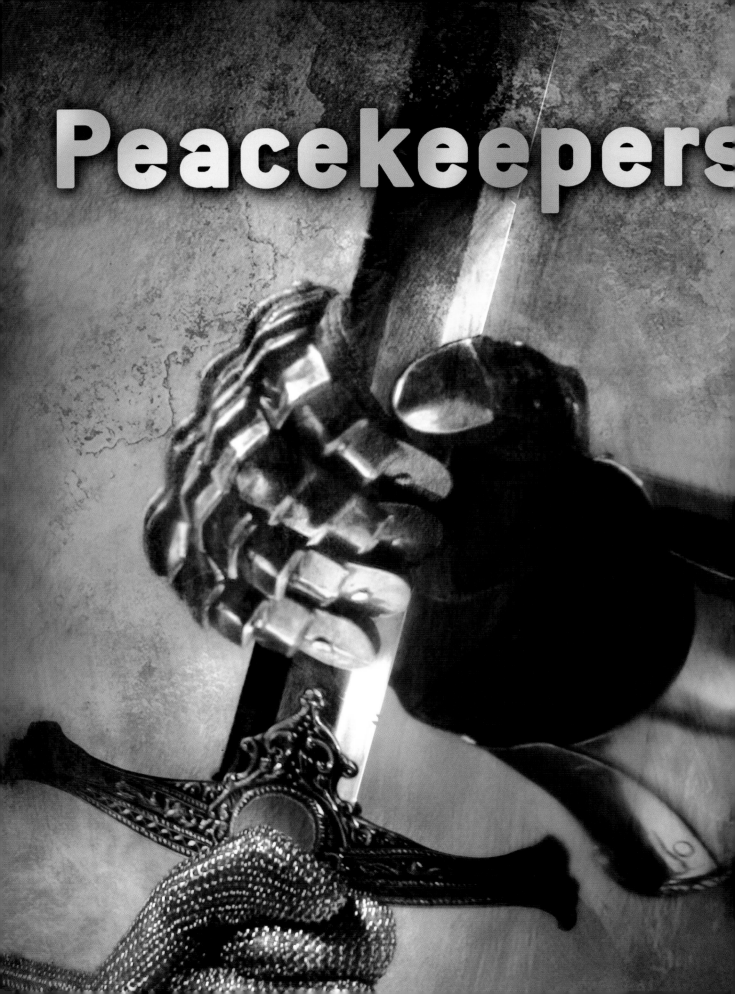

Peacekeepers

of the IMMUNE SYSTEM

Regulatory T cells, only recently proven to exist, keep the body's defenses from attacking the body itself. Manipulations of these cells could offer new treatments for conditions ranging from diabetes to organ rejection

By Zoltan Fehervari and Shimon Sakaguchi

"*Horror autotoxicus.*"

A century ago the visionary bacteriologist Paul Ehrlich aptly coined that term to describe an immune system attack against a person's own tissues. Ehrlich thought such autoimmunity—another term he coined—was biologically possible yet was somehow kept in check, but the medical community misconstrued his two-sided idea, believing instead that autoimmunity had to be inherently impossible. After all, what wrong turn of evolution would permit even the chance of horrendous, built-in self-destruction?

Slowly, though, a number of mysterious ailments came to be recognized as examples of horror autotoxicus—among them multiple sclerosis, insulin-dependent diabetes (the form that commonly strikes in youth) and rheumatoid arthritis. Investigators learned, too, that these diseases usually stem from the renegade actions of white blood cells known as CD4+ T lymphocytes (so named because they display a molecule called CD4 and mature in the thymus). Normal versions of these cells serve as officers in the immune system's armed forces, responsible for unleashing the system's combat troops against disease-causing microorganisms. But sometimes the cells turn against components of the body.

Ehrlich was correct in another way as well. Recent work has identified cells that apparently exist specifically to block aberrant immune behavior. Called regulatory T cells, they are a subpopulation of CD4+ T cells, and they are vital for maintaining an immune system in harmony with its host. Increasingly, immunologists are also realizing that these cells do much more than quash autoimmunity; they also influence the immune system's responses to infectious agents, cancer, organ transplants and pregnancy. We and others are working to understand exactly how these remarkable cells carry out their responsibilities and why they sometimes function imperfectly. The findings should reveal ways to regulate the regulators and thus to depress or enhance immune activity as needed and, in so doing, to better address some of today's foremost medical challenges.

Imperfect Defenses

LIKE THE IMMUNOLOGISTS of Ehrlich's time, many people today would be dismayed to know that no matter how healthy they may be, their bodies harbor potentially destructive immune system cells quite capable of triggering autoimmune disease. Yet this immunological sword of Damocles can be easily demonstrated. If a mouse, for example, is injected with proteins from its own central nervous system, along with an adjuvant (a generalized immune system stimulus), a destructive immune reaction ensues. Much as in multiple sclerosis, T cells launch an attack on the animal's brain and spinal cord.

By varying the source of the injected self-protein, researchers can provoke other autoimmune diseases in laboratory animals—which indicates that potentially harmful immune sys-

WHEN THE IMMUNE SYSTEM wields its weapons inappropriately, regulatory T cells, also called T-regs, restrain them.

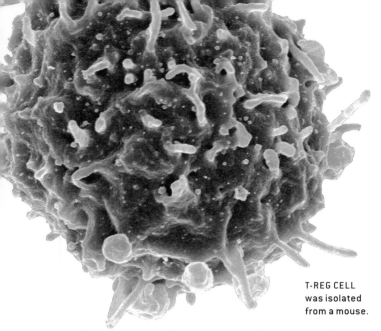

T-REG CELL was isolated from a mouse.

tem cells can mount self-attacks on a wide variety of tissues. The risk appears to hold true in humans, too, because autoreactive immune system cells can be captured readily from the blood of a healthy person. In a test tube, they react strongly to samples of that person's tissues.

Given such demonstrations of clear and imminent danger, investigators naturally wondered how it is that most animals and humans are untroubled by autoimmune disease. Put another way, they wanted to know how the immune system distinguishes threats such as microbes from a person's own tissues. They found that to achieve self-tolerance—the ability to refrain from attacking one's own organs—the immune system enlists numerous safeguards. The first defense, at least where T cells are concerned, occurs in the thymus, which lies inconspicuously in front of the heart. In the thymus, immature T cells undergo a strict "education" in which they are programmed to not react strongly (and therefore harmfully) to any bodily tissues. Disobedient cells are destroyed. No system is perfect, though, and in fact a small number of autoaggressive T cells slip through. Escaping into the bloodstream and into lymph vessels, they create the immune system's potential for unleashing autoimmune disease.

Blood and lymph vessels are where a second line of defense comes into play. This layer of protection against autoimmunity has several facets. Certain tissues, including those of the brain and spinal cord, are concealed from immune cell patrols simply by having a paucity of blood and lymph vessels that penetrate deep into the tissue. Their isolation, however, is not absolute, and at times, such as when the tissues are injured, self-reactive immune cells can find a way in. Additional modes of protection are more proactive. Immune cells showing an inappropriate interest in the body's own tissues can be targeted for destruction or rendered quiescent by other immune system components.

Among the immune cells that carry out these proactive roles, regulatory T cells may well be the most crucial. The majority, if not all of them, learn their "adult" roles within the thymus, as other T cells do, then go forth and persist throughout the body as a specialized T cell subpopulation.

Discovering the Peacekeepers

FINDINGS HINTING AT the existence of regulatory T cells date back surprisingly far. In 1969 Yasuaki Nishizuka and Teruyo Sakakura, working at the Aichi Cancer Center Research Institute in Nagoya, Japan, showed that removing the thymus from newborn female mice had a curious outcome: the animals lost their ovaries. At first it was thought that the thymus must secrete some kind of hormone needed for survival of the developing ovaries. Later, though, it turned out that immune system cells invaded the ovaries. The ovarian destruction was therefore an autoimmune disease, which had presumably been unleashed by the animals' loss of a countervailing regulatory process. If the mice were inoculated with normal T cells, the autoimmune disease was inhibited. T cells, then, could at times police themselves somehow.

In the early 1970s John Penhale of the University of Edinburgh made analogous observations in adult rats, and Richard Gershon of Yale University became the first to propose the existence of a T cell population capable of damping immune responses, including autoaggressive ones. This hypothetical immune system member was christened the suppressor T cell. At the time, though, no researcher was able to actually find one or pinpoint the molecular action by which one immune system cell could restrain another. Consequently, the concept of the suppressor T cell languished along the fringes of mainstream immunology.

Despite the negative atmosphere, some researchers persisted in trying to identify T cells with an ability to prevent autoimmune disease. The basic hope was to discover a telltale molecular feature at the surface of such cells—a "marker" by which suppressor T cells could be distinguished from other cells. Beginning in the mid-1980s, various candidate markers were explored.

In 1995 one of us (Sakaguchi) finally demonstrated that a molecule called CD25 was a reliable marker. When, in studies of mice, he removed CD4+ T cells displaying that molecule, organs such as the thyroid, stomach, gonads, pancreas and salivary glands came under an autoimmune attack character-

Overview/*Immune Regulators*

- For years, immunologists doubted that cells specifically responsible for suppressing immune activity existed. But they do. They are called regulatory T cells.
- These so-called T-reg cells combat autoimmunity. They also help the body resist repeat infections by a returning invader, protect needed bacteria in the gut and aid in sustaining pregnancy. On the negative side, they abet cancer cells in escaping immune attacks.
- Ongoing research promises to yield new therapies for autoimmune disorders and cancer and could lead to treatments that would spare organ transplant recipients from having to take immunosuppressive drugs for life.

T-reg cells help to ensure that immune system components— including T cells that fight infections—refrain from attacking normal tissues. The thymus, where all T cell varieties mature, directly eliminates many strongly autoreactive cells (*left*), but its vigilance is imperfect, so T-regs patrol the body in search of renegades (*right*).

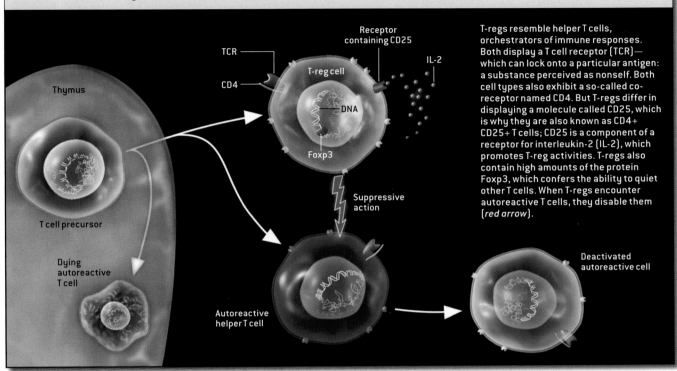

T-regs resemble helper T cells, orchestrators of immune responses. Both display a T cell receptor (TCR)— which can lock onto a particular antigen: a substance perceived as nonself. Both cell types also exhibit a so-called co-receptor named CD4. But T-regs differ in displaying a molecule called CD25, which is why they are also known as CD4+ CD25+ T cells; CD25 is a component of a receptor for interleukin-2 (IL-2), which promotes T-reg activities. T-regs also contain high amounts of the protein Foxp3, which confers the ability to quiet other T cells. When T-regs encounter autoreactive T cells, they disable them (*red arrow*).

ized by dramatic inflammation: white blood cells swarmed into the organs and damaged them.

In an important confirmatory experiment, T cell populations obtained from normal mice were depleted of their CD4+ CD25+ T cells, which evidently made up only a small proportion (at most, 10 percent) of the overall T cell pool. Then T cells left behind were transferred to mice engineered to lack an immune system of their own. This maneuver caused autoimmune disease. And the more complete the depletion was in the donor animals, the more severe the spectrum of disease became in the recipients—with comprehensive depletion often proving to be fatal. Reintroducing CD4+ CD25+ T cells, even in small numbers, conferred normal immunity and protected the animals from these disorders. Experiments conducted wholly in test tubes also produced valuable confirmatory evidence. Perhaps to absolve "suppressor cells" of any lingering stigma, immunologists started to call them CD25+ regulatory T cells, or simply T-regs.

How Do T-regs Work?

TO THIS DAY, the precise ways in which T-regs suppress autoimmune activity have remained mysterious, making their function a continuing subject of intense inquiry. The cells appear capable of suppressing a wide variety of immune system cells, impeding the cells' multiplication and also their other activities, such as secretion of cell-to-cell chemical signals

(cytokines). And researchers tend to agree that T-regs are activated by direct cell-to-cell contacts. Beyond that, the picture is rather murky [*see box on next two pages*].

Recently, however, our laboratory at Kyoto University and, independently, Alexander Rudensky's group at the University of Washington and Fred Ramsdell's group at CellTech R&D in Bothell, Wash., found a fresh clue as to how T-regs develop and function. The cells contain a large amount of an intracellular molecule called Foxp3. In fact, the enrichment is greater than has been reported for any other T-reg molecular feature.

Foxp3 is a transcription factor: a molecule that regulates the activity of specific genes, thereby controlling a cell's production of the protein that each such gene encodes. Because proteins are the main worker molecules in cells, altered production of one or more of them can affect how a cell functions. In the case of Foxp3, the changes it induces in gene activity apparently turn developing T cells into T-regs. Indeed, artificially introducing Foxp3 into otherwise unremarkable T cells provokes a reprogramming, by which the cells acquire all the suppressive abilities of full-fledged T-regs produced by the thymus. A type of mouse called the Scurfy strain, long known to researchers, has recently been found to have only an inactive, mutant form of the Foxp3 protein, along with a total absence of T-regs. The consequence is an immune system gone haywire, with massive inflammation

HOW DO T-REGS PREVENT AUTOIMMUNITY?

No one fully understands how T-regs block autoimmune attacks. Three reasonable possibilities follow. All three involve interfering with a key step in triggering immune responses: signaling between T cells and antigen-presenting cells (APCs). Before helper T cells will call forth other troops and

before "cytotoxic" T cells will attack tissue perceived to be infect◼ APCs must display antigens for the cells' perusal. If the T cell rece◼ (TCR) of a helper or cytotoxic cell recognizes a displayed antigen ◼ also receives certain other signals from the APC, the T cells will

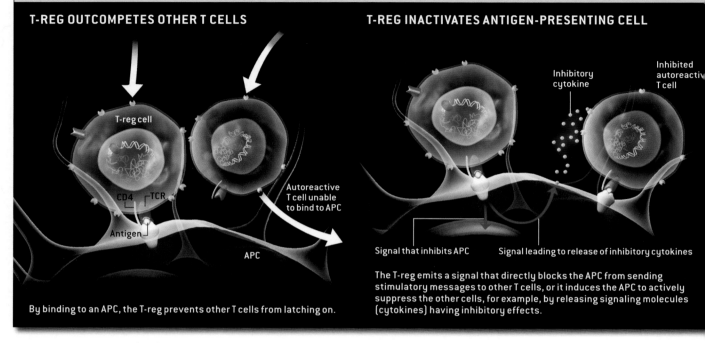

T-REG OUTCOMPETES OTHER T CELLS

By binding to an APC, the T-reg prevents other T cells from latching on.

T-REG INACTIVATES ANTIGEN-PRESENTING CELL

The T-reg emits a signal that directly blocks the APC from sending stimulatory messages to other T cells, or it induces the APC to actively suppress the other cells, for example, by releasing signaling molecules (cytokines) having inhibitory effects.

in numerous organs, leading to the animals' early death.

Of course, investigators study T-regs in animals such as mice so that the knowledge gained may be applied to humans. So what evidence is there that T-regs are indeed important in humans—or that they exist in us at all?

It turns out that the molecular features characteristic of T-regs in rodents are also characteristic of a subset of T cells in humans. In humans, as in rodents, these cells exhibit the CD25 molecule and have a high content of Foxp3. In addition, the cells are immunosuppressive, at least in a test tube.

Perhaps the most compelling indications that they are vital to human health come from a rare genetic abnormality called IPEX (*i*mmune dysregulation, *p*olyendocrinopathy, *e*nteropathy, *X*-linked syndrome). Arising from mutations in a gene on the X chromosome, IPEX affects male children, who unlike females inherit only one X chromosome and hence have no chance of inheriting a second, normal copy of the gene, which would encode a healthy version of the affected protein. In males the mutation results in autoimmune disease affecting multiple organs, including the thyroid and (as happens in insulin-dependent diabetes) the pancreas, and also in chronic intestinal inflammation (inflammatory bowel disease) and uncontrolled allergy (food allergy and severe skin inflammation), all of which can be understood as varied manifestations of the hyperactivity of an immune system unrestrained by T-regs. Death comes in infancy or soon after, with contributing causes ranging from autoimmune diabetes to severe diarrhea. The

specific genetic flaw underlying IPEX has recently proved to be mutation in none other than *Foxp3*. IPEX is therefore the human counterpart of the illness in Scurfy mice.

Beyond Self-Tolerance

THE EVIDENCE, THEN, indicates that T-regs do prevent autoimmune disease in humans. But the cells also appear to serve health in other ways, including participating (in some surprising ways) in responses to microbes.

Throughout the 1990s Fiona Powrie and her colleagues at the DNAX Research Institute in Palo Alto, Calif., experimented with transferring T cell populations depleted of T-regs into mice engineered to lack an immune system of their own. In one set of studies, the transfer induced a severe, often fatal form of inflammatory bowel disease. But the aberrant immune activity was not directed primarily at bowel tissue itself.

The bowels of rodents, like those of humans, are home to a vast bacterial population, typically more than a trillion for every gram of intestinal tissue. Although these bacteria are foreign, they are usually far from harmful; indeed, they promote the digestion of food and even displace dangerous bacteria, such as salmonella, that would otherwise try to colonize the intestines. Normally the immune system tolerates the presence of the helpful population. But in Powrie's mice, it attacked. And in doing so, the transplanted immune cells caused collateral damage to the recipient's gut. Yet transfer of T-regs caused no problems. In fact, if the T-regs were transferred

ome active against the bearer of that antigen—even if the antigen is
the body itself, instead of from an infectious agent. The TCRs of T-regs
recognize particular antigens, and they specifically suppress T cells
focus on those same antigens.

EG QUIETS OTHER T CELLS DIRECTLY

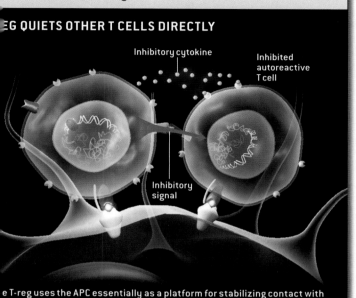

Inhibitory cytokine

Inhibited
autoreactive
T cell

Inhibitory
signal

e T-reg uses the APC essentially as a platform for stabilizing contact with
other T cell bound to the APC. Then the T-reg sends an inhibitory signal
ectly into the T cell or emits inhibitory molecules that act at close range.

along with the other T cells, they prevented the bowel disease that would otherwise have ensued. Overall, the immune system appeared to be on a hair trigger, prepared to assault gut bacteria and held in check only by T-regs.

A similar hair trigger may affect the immune system's responses to harmful foreigners. On the one hand, T-regs might rein in an overemphatic response. On the other hand, the reining in might keep an invader from being totally destroyed, enabling it to persist and potentially flare up again. For example, some findings suggest that failure to clear the stomach of a bacterium called *Helicobacter pylori,* now known to cause stomach ulcer, stems from blunting by T-regs of the immune system's weaponry.

Work by David Sacks and his colleagues at the National Institutes of Health has revealed further complexity. It implies that leaving a few survivors among invading organisms may not be entirely a bad thing. The researchers infected mice with a fairly innocuous parasite. Even when the immune system was fully intact, it allowed a small number of parasites to remain, after which reinfection triggered a prompt, efficient response. If the immune system was depleted of its T-regs, however, the parasite was completely purged, but reinfection was dealt with inefficiently, as if the mice had never before encountered the invader. Hence, T-regs appear to contribute to maintaining immunological memory, a process that is crucial for immunity to repeated infection and that also underlies the success of vaccination.

Research hints, too, at a role for T-regs in protecting pregnancies. Every pregnancy unavoidably poses quite a challenge to the mother's immune defenses. Because the fetus inherits half its genes from the father, it is genetically half-distinct from its mother and thus is in essence an organ transplant. Within the trophoblast, the placental tissue that attaches the fetus to the uterine wall, a number of mechanisms give the fetus some safety from what would amount to transplant rejection. The trophoblast not only presents a physical barrier to would-be attackers in the mother's blood but also produces immunosuppressive molecules.

The mother's immune system seems to undergo changes as well. Reports of women in whom an autoimmune disease such as multiple sclerosis abates during pregnancy provide anecdotal evidence that T-regs become more active. Some recent experiments offer more direct support. At the University of Cambridge, Alexander Betz and his colleagues have shown that during pregnancy in mice, maternal T-regs expand in number. Conversely, an experimentally engineered absence of T-regs leads to fetal rejection marked by a massive infiltration of immune cells across the maternal-fetal boundary. It is tempting to speculate that in some women, insufficient T-reg activity may underlie recurrences of spontaneous abortion.

Recruiting the Regulators

IN T-REGS, nature clearly has crafted a potent means of controlling immune responses. Tapping into this control would make T-regs a potentially powerful therapeutic ally against a wide range of medical disorders. It is still too early to expect to see applications in doctors' offices, but the available data suggest that delivering T-regs themselves, or perhaps medicines that increase or decrease their activity, could provide novel treatments for a variety of conditions. Indeed, some human trials are under way.

The most obvious application would involve enhancing T-reg activity to fight autoimmune diseases, and drug therapy is being explored in patients with multiple sclerosis and psoriasis, among other conditions. Pumping up T-reg activity might also be useful for treating allergies. The ease with which T-regs can keep immune responses at bay suggests that T-reg-based therapies could hold particular promise for preventing rejection of transplanted organs. The ideal would be for transplant recipients to tolerate grafts as well as they do their own tissues. Also ideal would be a tolerance that

THE AUTHORS

ZOLTAN FEHERVARI and *SHIMON SAKAGUCHI* began collaborating in 2002, when Fehervari took a postdoctoral position in Sakaguchi's laboratory at the Institute for Frontier Medical Sciences of Kyoto University in Japan. Fehervari is now a research associate in the department of pathology at the University of Cambridge, where he earned a Ph.D. in immunology. Sakaguchi is professor and chair of the department of experimental pathology at Kyoto. He began searching for regulatory T cells in the early 1980s and has studied them ever since.

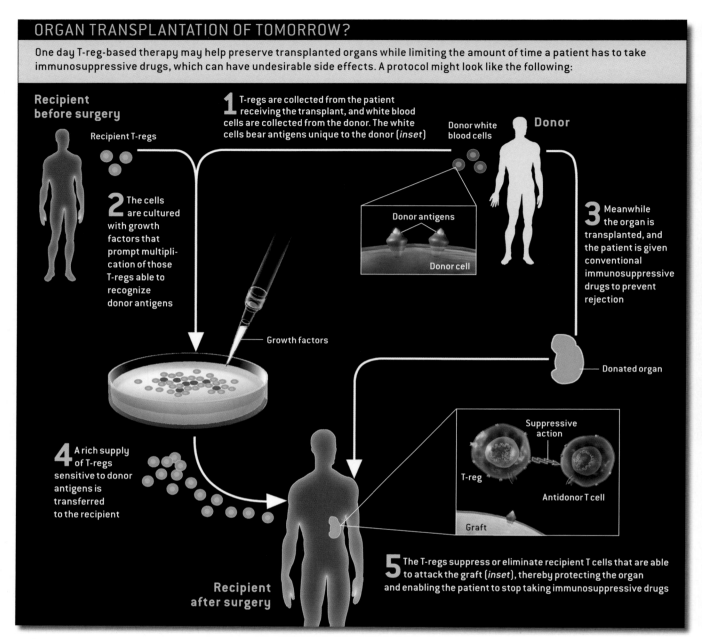

ORGAN TRANSPLANTATION OF TOMORROW?

One day T-reg-based therapy may help preserve transplanted organs while limiting the amount of time a patient has to take immunosuppressive drugs, which can have undesirable side effects. A protocol might look like the following:

Recipient before surgery

Recipient T-regs

1 T-regs are collected from the patient receiving the transplant, and white blood cells are collected from the donor. The white cells bear antigens unique to the donor (*inset*)

Donor white blood cells

Donor

2 The cells are cultured with growth factors that prompt multiplication of those T-regs able to recognize donor antigens

Donor antigens

Donor cell

3 Meanwhile the organ is transplanted, and the patient is given conventional immunosuppressive drugs to prevent rejection

Growth factors

Donated organ

4 A rich supply of T-regs sensitive to donor antigens is transferred to the recipient

Suppressive action

T-reg

Antidonor T cell

Graft

Recipient after surgery

5 The T-regs suppress or eliminate recipient T cells that are able to attack the graft (*inset*), thereby protecting the organ and enabling the patient to stop taking immunosuppressive drugs

---

endures as a permanent state of affairs, without need for immunosuppressive drugs, which can have many side effects.

The opposite type of T-reg-based therapy would be a selective depletion of T-regs to counter unwanted immunosuppression and, consequently, to strengthen beneficial immune responses. In practice, a partial depletion might be preferred to a complete one, because it should pose less risk of inducing autoimmune disease. Best of all would be removal solely of those T-regs that were specifically blocking a useful immune response. The depletion strategy might be especially advantageous against infectious diseases that the immune system, left to itself, tends to combat inadequately—perhaps tuberculosis or even AIDS.

In addition, T-reg reduction might be advantageous for fighting cancer. Much evidence suggests that circulating immune cells keep a lookout for molecular aberrations that occur as a cell becomes cancerous. To the extent that T-regs impede this surveillance, they might inadvertently help a malignancy take root and grow. In fact, some cancers appear to encourage such help: they secrete molecular signals capable of attracting T-regs and of converting non-T-regs into T-regs. Some findings suggest, for example, that cancer patients have abnormally high numbers of active T-regs both in their blood and in the tumors themselves. Much of today's research into therapeutic manipulations of T-regs focuses on cancer.

Technical Challenges

SO FAR INVESTIGATORS are finding it challenging to develop medicines able to deplete or expand T-reg populations within a patient's body. To be most useful, these drugs would

Some T-Reg-Based Therapies under Study

The therapies listed below are among those in or likely to enter human trials. Most of the drugs under study aim to deplete or inhibit T-reg cells, so as to increase antitumor immune responses normally tempered by the cells. Delivery of such agents into the body would need to be managed carefully, though, to ensure that reducing T-reg activity does not lead to autoimmunity.

EFFECT ON T-REGS	EXAMPLES OF DISORDERS BEING TARGETED	TREATMENT APPROACHES
Depletion or inhibition (to enhance immunity)	Cancers of the skin (melanoma), ovary, kidney	A toxin fused to a substance, such as interleukin-2, able to deliver the toxin to T-regs
		Monoclonal antibodies (which bind to specific molecules) that have shown an ability to induce T-reg death or to block the cells' migration into tumors
Multiplication in patient (to dampen autoimmunity)	Multiple sclerosis, psoriasis, Crohn's disease, insulin-dependent diabetes	Vaccine composed of T cell receptor constituents thought to stimulate T-reg proliferation
		A monoclonal antibody that appears to stimulate T-regs by binding to a molecule called CD3
Multiplication in the lab, for delivery to patient	Graft-versus-host disease (immune cells in donated bone marrow attack recipient tissue)	Culture donor T-regs with selected antibodies and growth factors, then deliver resulting T-reg population before or at the time of bone marrow transplant (for prevention) or if graft-versus-host disease arises

usually need to act on the subsets of T-regs that have roles in a particular disorder, yet scientists often do not know precisely which T-regs to target.

Devising therapies based on administering T-regs themselves is difficult as well. One of the main obstacles is the need to obtain enough of the cells. Although researchers have found that T-regs can operate at low abundance relative to the cells they are suppressing, control of a human autoimmune disease would probably require tens of millions of T-regs. Acquiring such numbers of these relatively rare cells from a person's circulation might be impossible. Accordingly, some technique to expand their numbers outside the body would seem to be imperative.

Luckily, it also seems that this numbers game can be won. Worldwide, several research groups have reported that cells with immunosuppressive actions can be generated in relatively large numbers by treating ordinary T cells with a well-defined "cocktail" of biochemical signals. Whether the engendered cells, termed Tr1 cells, are identical to T-regs remains unclear, but it is beyond dispute that the cells are profoundly immunosuppressive.

Now that Foxp3 is known to be a key molecule controlling the development and function of T-regs, investigators may also be able to tailor-make large numbers of regulatory cells by using fairly standard laboratory techniques to transfer the *Foxp3* gene into more prevalent, and thus more easily obtainable, types of T cells. We and others are pursuing this approach intently and are also trying to identify the molecular events that switch on Foxp3 production during T-reg development. This knowledge might enable pharmaceutical researchers to fashion drugs specifically for that purpose, so that processing of cells outside the body and then infusing them would not be necessary.

For organ transplant patients, another way to obtain useful T-regs is under consideration. The procedure would involve removing T-regs from a prospective transplant recipient and culturing them with cells from the organ donor in a way that causes the T-regs most capable of suppressing rejection to multiply [see box on opposite page]. In rodents, T-regs generated in this manner have worked well. One of us (Sakaguchi) has shown, for example, that injection of a single dose of such T-regs at the time of skin grafting results in the graft's permanent acceptance, even though transplanted skin typically is rejected strongly. Meanwhile the treatment left the rest of the immune system intact and ready to fend off microbial invaders. The abundant research into T-regs suggests that such an approach can become a reality for humans and could be used to protect new transplant recipients until medications able to produce the same benefit more simply are developed.

Over the past decade, researchers' understanding of the immune system and how it governs its own actions has changed profoundly. In particular, it is now recognized that although the system permits potentially autodestructive T cells to circulate, it also deploys T cells capable of controlling them. Knowledge of how they develop and how they perform their remarkable immunosuppressive activities will be key in recruiting them for use against a host of debilitating and even fatal disorders. In permitting destruction of nonself while preventing destruction of self, T-regs may prove to be the ultimate immunological peacekeepers. [SA]

MORE TO EXPLORE

Naturally Arising CD4+ Regulatory T Cells for Immunologic Self-Tolerance and Negative Control of Immune Responses. Shimon Sakaguchi in *Annual Review of Immunology*, Vol. 22, pages 531–562; 2004.

Regulatory T-Cell Therapy: Is It Ready for the Clinic? J. A. Bluestone in *Nature Reviews Immunology*, Vol. 5, No. 4, pages 343–349; April 2005.

Regulatory T Cells, Tumour Immunity and Immunotherapy. Weiping Zou in *Nature Reviews Immunology*, Vol. 6, No. 4, pages 295–307; April 2006.

T Lymphocytes: Regulatory. Zoltan Fehervari and Shimon Sakaguchi in *Encyclopedia of Life Sciences*. Wiley InterScience, 2006. Available at www.els.net

Questions for Review

"Peacekeepers of the Immune System"
by Zoltan Fehervari and Shimon Sakaguchi

TESTING YOUR COMPREHENSION

1. Which one of the following does not belong with the others?
   a. Insulin-dependent diabetes
   b. Multiple sclerosis
   c. Rheumatoid arthritis
   d. Ulcers

2. Autoimmune diseases were
   a. discovered in the 21st century.
   b. discovered in the 20th century.
   c. unknown before 1950.
   d. discovered by Pasteur.

3. CD25 identifies
   a. B cells.
   b. cytotoxic T cells.
   c. helper T cells.
   d. regulatory T cells.

4. T-regs "learn" their role
   a. in the intestine.
   b. in the thymus.
   c. during adulthood.
   d. in the blood.

5. Foxp3 activates genes causing a cell to
   a. become a T cell.
   b. produce cytokines.
   c. become a T-reg cell.
   d. produce antibodies.

6. What happens if a mouse is injected with its own central nervous system (CNS) proteins?
   a. Nothing, the CNS proteins are self proteins.
   b. The mouse produces antibodies against foreign CNS proteins.
   c. The mouse produces antibodies against its own CNS tissues.
   d. T-regs destroy the mouse's CNS.

7. In mice lacking T-regs, inflammatory bowel disease occurred because
   a. the immune system killed normal intestinal microbiota.
   b. normal intestinal microbiota caused a fatal infection.
   c. a memory response occurred.
   d. antibodies were not made.

8. Which one of the following does not belong with the others?
   a. Tr1
   b. Inhibiting antigen-presenting cells
   c. Culturing recipient T-regs with donor cells
   d. Transferring Foxp3 to T cells

9. In pregnant mice, T-regs appear to
   a. prevent fetal rejection.
   b. cause spontaneous abortion.
   c. cause an immune response to the fetus.
   d. cause multiple sclerosis.

10. Immune dysregulation, polyendocrinopathy, enteropathy, X-linked syndrome (IPEX) is a severe autoimmune disease resulting in death during infancy. IPEX is due to a mutation in the gene encoding
    a. antibodies.
    b. cytokines.
    c. Foxp3.
    d. growth factors.

MICROBIOLOGY IN SOCIETY

1. Assume that injection of T-regs could cure multiple sclerosis, insulin-dependent diabetes, and rheumatoid arthritis. Research on T-regs is in its infancy and it could cost billions of dollars to get to pharmaceutical T-regs to treat these diseases. Who should pay for this research: government, universities, pharmaceutical companies?

2. Assume that pharmaceutical company X has developed a drug to treat an autoimmune disease using recombinant T-regs (rT-reg). These rT-regs were made by inserting the Foxp3 gene into stem cells. Should these recombinant stem cells be used?

3. Assume that T-reg therapy works and that treatment costs $1,000 per person. Should every person who needs it be guaranteed to receive the treatment? If so, who's paying? If not, who decides who gets treatment?

THINKING ABOUT MICROBIOLOGY

1. In 1905, Paul Ehrlich thought "horror autotoxicus" was possible but normally kept in check. What lab criteria would characterize "autotoxicus" and how does the body prevent "horror autotoxicus?"

2. The immune system is designed to attack chemicals that are not the body's own tissues, such as pathogens and genetically nonidentical organ transplants, so why does the maternal immune system not attack a developing fetus?

3. In mice, elimination of T-regs before transplanting a tumor resulted in tumor-free survival in 70% of mice. Depletion of T-regs in animals with established tumors had no therapeutic effect. What can you conclude about the use of T-regs in cancer therapy?

WRITING ABOUT MICROBIOLOGY

1. Describe three mechanisms by which T-regs might block autoimmune attacks.

2. Discuss an example of the importance of microorganisms to the immune system.

3. In the table of therapies on page 19, a suggested cancer treatment is to deliver a toxin to T-regs. What is the value of killing T-regs?

Answers can be found on The Microbiology Place website. Go to www.microbiologyplace.com, click on the cover of your textbook, and type in your login name and password (using the access code found in the front pages of your textbook). Then, click on Current Issues Magazine Answers on the left navigation bar.

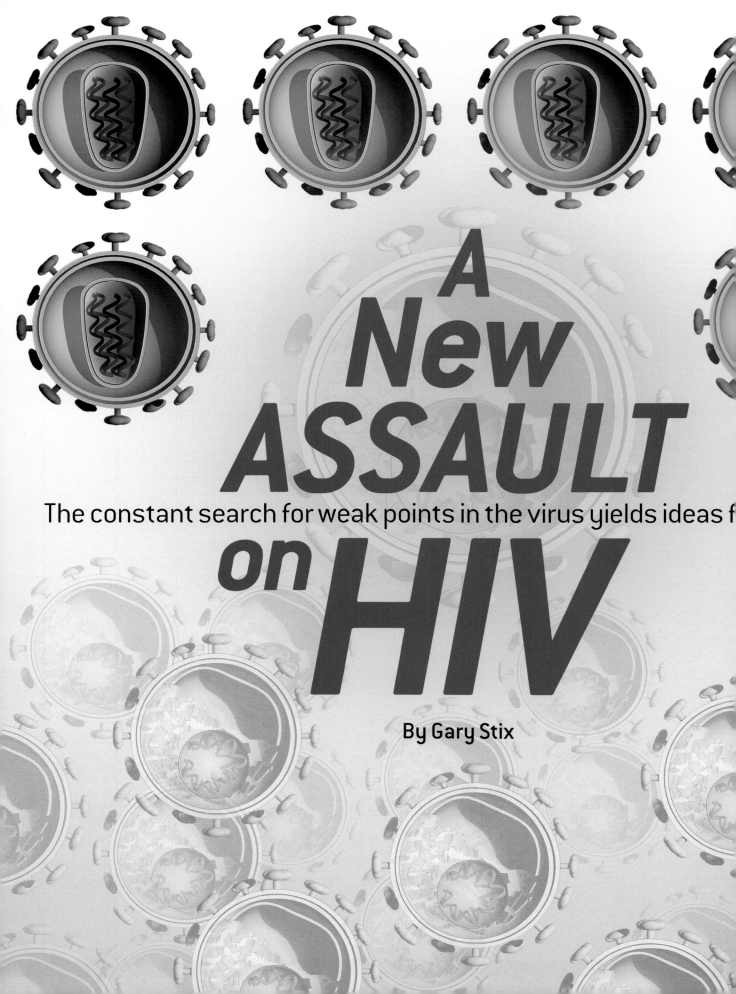

A New ASSAULT on HIV

The constant search for weak points in the virus yields ideas f

By Gary Stix

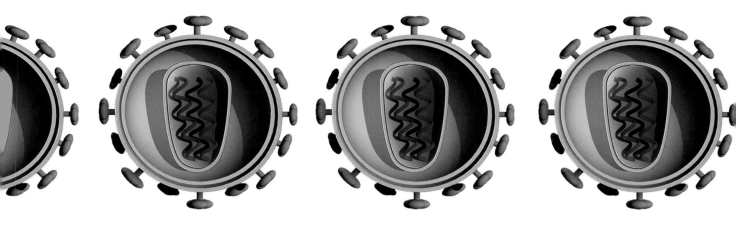

The field of virology spends a substantial chunk of its resources inspecting every minute step of the HIV life cycle—from the binding and entry of the virus into an immune cell to its replication and release of a new virus from the host cell and, finally, the seeking of a new cell on which to prey. The last major new class of anti-HIV drugs emerged about a decade ago with the introduction of the protease inhibitors, which curb the action of an enzyme that is critical to a late stage of viral replication.

At the time, a few members of the HIV research community wondered whether protease inhibitors could provide the basis for a cure. The ingenuity of the virus has proved the hollowness of that hope. As many as half of HIV-positive patients under treatment in the U.S. were found in one study to be infected with viruses that have developed resistance to at least one of the drugs in their regimen. Clinicians can choose from more than 20 pharmaceuticals among protease inhibitors and two classes of drug that prevent the invading virus from copying its RNA into DNA, thereby sabotaging viral replication. Combinations of these agents are administered to counteract the virus's inherent mutability, but that strategy does not always ward off resistance to the medicines, including the protease inhibitors. "Given increasing resistance to protease inhibitors, it's of paramount importance to identify new ways to interfere with the virus replication cycle," asserts Eric Freed, a researcher in the HIV drug resistance program at the National Institutes of Health.

Drugs that interrupt the beginning, middle and end of viral processing within the host are now in various stages of development. Academic researchers and Panacos, a small biotechnology outfit based in Watertown, Mass., are taking inspiration from the success of protease inhibitors by developing drug candidates known as maturation inhibitors that block protease activity in a novel way. Protease inhibitors mount a direct attack on the HIV protease, preventing the enzyme from processing a viral protein called GAG. When GAG proteins are cut properly, pieces spliced out of it form the conical protective core, or capsid, that encloses RNA. In contrast, the Panacos maturation inhibitor blocks a site on the GAG protein where the protease normally binds, keeping the protease from clipping GAG correctly. As a result, the capsid does

wholly new class of drug

not form appropriately and the virus cannot infect another cell.

Looking for Leads

THE PATH TO THE PANACOS drug candidate began in the mid-1990s, when the company Boston Biomedica undertook a collaboration with a professor from the University of North Carolina at Chapel Hill to screen compounds from a collection of traditional Chinese herbs for biochemical activity against HIV. Kuo-Hsiung Lee's laboratory turned up a potential drug lead in a Taiwanese herb.

The substance, betulinic acid, had weak activity against HIV. After the lab separated the compound into its chemical constituents, the investigators found that one of these components, when chemically modified, exhibited a much stronger effect. "Betulinic acid had activity against HIV at the micromolar level," says Graham Allaway, Panacos's chief operating officer. "This derivative had activity at the nanomolar level."

Six years ago Boston Biomedica spun off its HIV research unit into Panacos, which began to investigate the compound, by then named PA-457. PA-457 was not just another Taxol, the anticancer drug that required the controversial felling of rare yew trees until a semisynthetic substitute was found. Panacos did not need a steady source of Taiwanese herbs. Betulinic acid could be extracted from ubiquitous plane and birch trees, and a subsequent processing step yielded the desired molecule.

Even though researchers understood that PA-457 seemed to have activity against all strains of HIV, they needed to find out how the betulinic acid derivative worked against the virus on the molecular level. The company wanted a new class of drug, not just another protease inhibitor. It contacted Freed's laboratory at the NIH, which studies the virus life cycle.

Freed's group and Panacos determined that the drug worked late in the viral replication process, apparently at the stage of capsid formation. The researchers already knew that the HIV capsid forms when newly made GAG at-

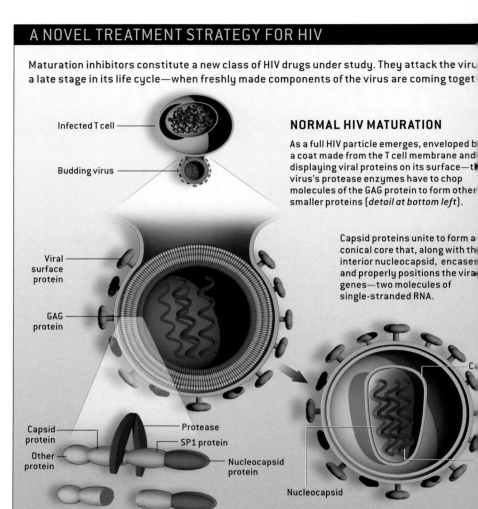

A NOVEL TREATMENT STRATEGY FOR HIV

Maturation inhibitors constitute a new class of HIV drugs under study. They attack the viru[s] a late stage in its life cycle—when freshly made components of the virus are coming toget[her]

Infected T cell

Budding virus

NORMAL HIV MATURATION

As a full HIV particle emerges, enveloped b[y] a coat made from the T cell membrane and displaying viral proteins on its surface—th[e] virus's protease enzymes have to chop molecules of the GAG protein to form other smaller proteins (*detail at bottom left*).

Capsid proteins unite to form a conical core that, along with the interior nucleocapsid, encases and properly positions the viral genes—two molecules of single-stranded RNA.

Viral surface protein

GAG protein

Capsid protein

Other protein

Protease

SP1 protein

Nucleocapsid protein

Nucleocapsid

Normal HIV particle

taches from inside the host T cell to the cell's membrane and is then chopped by the HIV protease into smaller pieces. They knew as well, from the development of protease inhibitors, that any disruption to the processing of GAG would cause the virus to become noninfectious. So they began to study PA-457's interaction with GAG to see exactly how it bollixed up the cutting of GAG into its requisite parts.

Cultivation of Resistance

TO UNDERSTAND HOW a compound works, scientists often begin by creating resistance, which lets them pin down the exact spot where the drug interacts with its target. To nurture resistance, Freed and his colleagues administered low doses of PA-457 to HIV-infected T cells

in culture. The genome of the resistant viruses was sequenced and compared with that of viruses that succumbed to the drug. That analysis located the site on the newly produced viruses that changed in the resistant versions. It turned out to be a site on GAG where the protease binds, and this alteration prevented PA-457 from blocking the enzyme's activity.

Analyzing the resistant strains allowed the researchers to ascertain that PA-457 was not simply another protease inhibitor. Most drugs, not just HIV inhibitors, work by tinkering with enzymes. "Targeting the substrate [instead of the enzyme] was unknown and surprising," Allaway comments. "As a result, we believe we will have a fairly strong patent position."

ew infectious particles that are beginning to "bud" from one infected T cell so that they
move on to infect another cell.

TREATED VIRUS

The drug candidate PA-457 works by attaching to the GAG
protein and preventing the protease from separating the capsid
protein from its neighbor in GAG—the SP1 protein (*detail*).

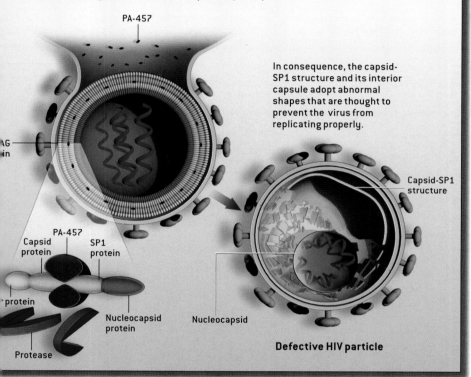

PA-457

In consequence, the capsid-
SP1 structure and its interior
capsule adopt abnormal
shapes that are thought to
prevent the virus from
replicating properly.

AG
in

PA-457

Capsid
protein

SP1
protein

protein

Capsid-SP1
structure

Nucleocapsid
protein

Nucleocapsid

Protease

Defective HIV particle

Cultivating resistant strains does not necessarily mean that the drug will have a limited therapeutic life span. In fact, resistance to PA-457 may not develop quickly, because the site where it binds on the GAG protein does not readily change from one HIV strain to another through mutations.

PA-457 has already passed through a midphase clinical trial that checked for drug activity in patients who took it for 10 days while another group received a placebo. HIV replicates so rapidly that a short trial can be used to determine whether a drug is attacking the pathogen in the body. Viral levels averaged a drop of 92 percent at the highest dose of 200 milligrams. The study looked for a decrease in so-called viral load of at least 70 percent as a preliminary sign of

the drug's effectiveness. Some patients, however, did not respond—and the company will determine in the next phase of testing whether it can give higher doses. "The main message is that this is an active drug and research should go forward," says Jeffrey M. Jacobson, chief of infectious diseases at Drexel University College of Medicine, who is the lead researcher in the clinical trials.

During the next round, investigators will be looking for interactions with other drugs, an essential test of any HIV drug prospect, because no treatment consists of a single drug therapy, given the threat of resistance. The Food and Drug Administration is encouraging tests earlier during clinical trials these days. In developing new HIV drugs, re-

searchers have at times detected these interactions only much later in the clinical trial process. If all goes according to plan, Panacos could file its final FDA approval application by 2008.

Other Immaturity Preservers

PA-457 IS NOT THE ONLY example of a maturation inhibitor, although it has progressed the furthest toward commercialization. At the University of Alabama and the University of Maryland, researchers working independently have identified small organic molecules that prevent the multitude of capsid subunits from joining up to form the finished casing. "We're trying to jam the parts so they don't fit together," says Peter Prevelige, a professor in the department of microbiology at the University of Alabama.

This strategy goes along with other approaches under development to sabotage the viral life cycle. Entry inhibitors, including one Panacos is working on, prevent the virus from entering the cell. (One injectable entry inhibitor has already received FDA approval, but the Panacos drug would be taken orally.) Among the other classes of drugs that have reached late-stage trials are integrase inhibitors, which undermine an enzyme that allows the viral-made DNA to integrate into the host DNA to produce new viral RNA. All these biological agents are needed—and more. Absent a vaccine—not a near-term prospect—the lowly virus, a nanometer-scale capsule of single-stranded RNA, will continue to outwit the best ideas that molecular biologists conjure. ⬛

MORE TO EXPLORE

PA-457: A Potent HIV Inhibitor That Disrupts Core Condensation by Targeting a Late Step in Gag Processing. F. Li et al. in *Proceedings of the National Academy of Sciences USA,* Vol. 100, No. 23, pages 13555–13560; November 11, 2003.

The Prevalence of Antiretroviral Drug Resistance in the United States. Douglas D. Richman et al. in *AIDS,* Vol. 18, No. 10, pages 1393–1401; July 2, 2004.

The Discovery of a Class of Novel HIV-1 Maturation Inhibitors and Their Potential in the Therapy of HIV. Donglei Yu et al. in *Expert Opinion on Investigational Drugs,* Vol. 14, No. 6, pages 681–693; June 2005.

Questions for Review

"A New Assault on HIV"

by Gary Stix

TESTING YOUR COMPREHENSION

1. What percentage of HIV patients has drug-resistant HIV infections?
    a. 25%
    b. 50%
    c. 75%
    d. 100%

2. Which of these is not a new class of antiretroviral drug?
    a. Entry inhibitor
    b. Integrase inhibitor
    c. Maturation inhibitor
    d. Protease inhibitor

3. Betulinic acid differs from a protease inhibitor because betulinic acid combines with
    a. the GAG protein.
    b. protease.
    c. viral capsids.
    d. viral DNA.

4. The site where betulinic acid binds
    a. is the same in every HIV strain.
    b. readily mutates.
    c. occurs in all viruses.
    d. is found in mature virions.

5. A maturation inhibitor could prevent
    a. viral entry into a cell.
    b. integration of viral DNA into a cell's chromosome.
    c. capsid formation.
    d. RNA synthesis.

6. In a clinical trial, betulinic acid caused
    a. a 92% increase in viral load.
    b. a 92% decrease in viral load.
    c. an 8% increase in viral load.
    d. an 8% decrease in viral load.

7. GAG is a(an)
    a. viral capsid.
    b. trick, hoax, or practical joke.
    c. protein made by HIV.
    d. instance of choking.

8. Betulinic acid had weak activity against HIV, however, after chemically modifying one component, activity changed from micromolar to nanomolar levels. This is a(an)
    a. decrease of 10 times.
    b. decrease of 100 times.
    c. increase of 1000 times.
    d. increase of 10,000 times.

9. HIV contains
    a. two molecules of single-stranded RNA.
    b. double-stranded RNA.
    c. single-stranded DNA.
    d. double-stranded DNA.

10. A fully mature HIV particle has an envelope made of
    a. protein.
    b. GAG.
    c. T-cell membrane.
    d. nucleoprotein.

MICROBIOLOGY IN SOCIETY

1. After laboratory testing, the effectiveness of a new drug in humans must be determined. Recently a pharmaceutical company made headlines when they tested an AIDS drug on orphans who contracted HIV at birth. Who should the subjects be in each phase of clinical trials?

   Phase 1: To test the safety, dosage levels, and response to a new treatment.

   Phase 2: To test whether a new treatment decreases the viral load.

   Phase 3: To compare the results of people taking a new treatment with results of people taking a standard treatment or no treatment to determine, for example, which group has better survival rates or fewer side-effects. (This usually involves a large number of people.)

2. Assume betulinic acid works quite well but there is a limited supply for the foreseeable future. Should it be made available to intravenous drug users? To incarcerated felons? To people in underdeveloped countries?

3. Current antiretroviral drugs can alter metabolism resulting in heart, liver, and kidney damage. What degree of side effects is acceptable?

THINKING ABOUT MICROBIOLOGY

1. Betulinic acid was identified by screening traditional Chinese herbs. Does this mean that all herbal remedies work to treat the disease for which they are prescribed?

2. Why is it important to look for a new class of drug, rather than a new drug that acts in a known manner?

3. Freed and his colleague cultured HIV-infected T cells in low doses of PA-457 to selected resistant viruses. What was the purpose of the T cells? How does this process select resistant viruses?

WRITING ABOUT MICROBIOLOGY

1. What does HIV protease do? How do protease inhibitors stop this enzyme?

2. Differentiate between a maturation inhibitor, protease inhibitor, integrase inhibitor, and entry inhibitor.

3. Why is it important to be able to find a drug in a "ubiquitous birch tree?"

Answers can be found on The Microbiology Place website. Go to www.microbiologyplace.com, click on the cover of your textbook, and type in your login name and password (using the access code found in the front pages of your textbook). Then, click on Current Issues Magazine Answers on the left navigation bar.

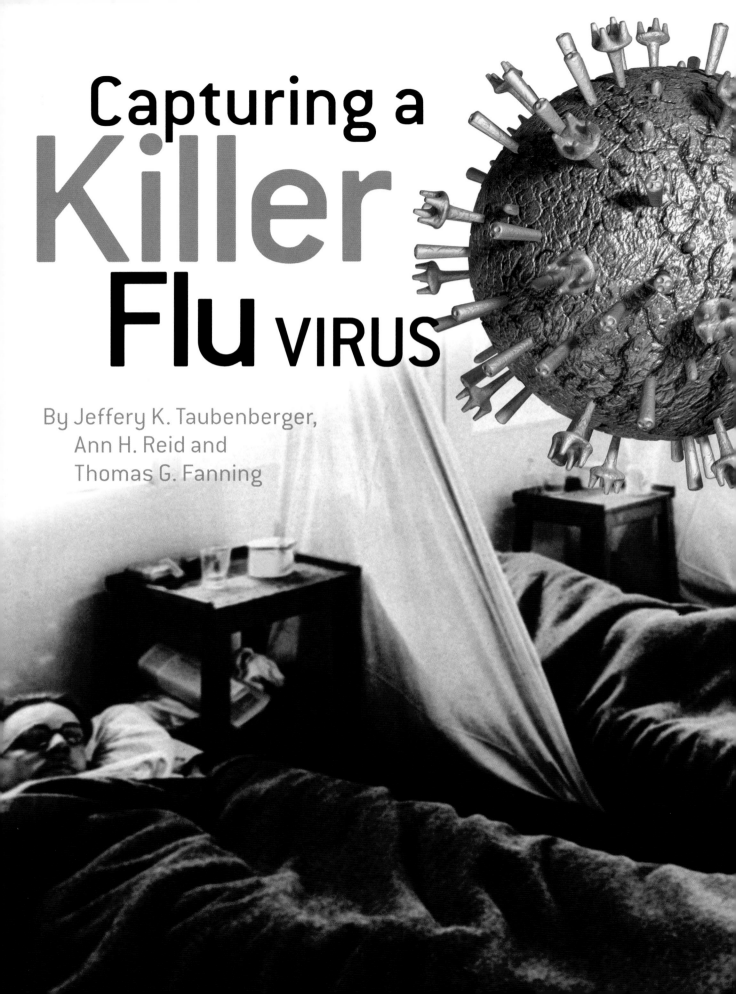

Capturing a
Killer
Flu VIRUS

By Jeffery K. Taubenberger,
Ann H. Reid and
Thomas G. Fanning

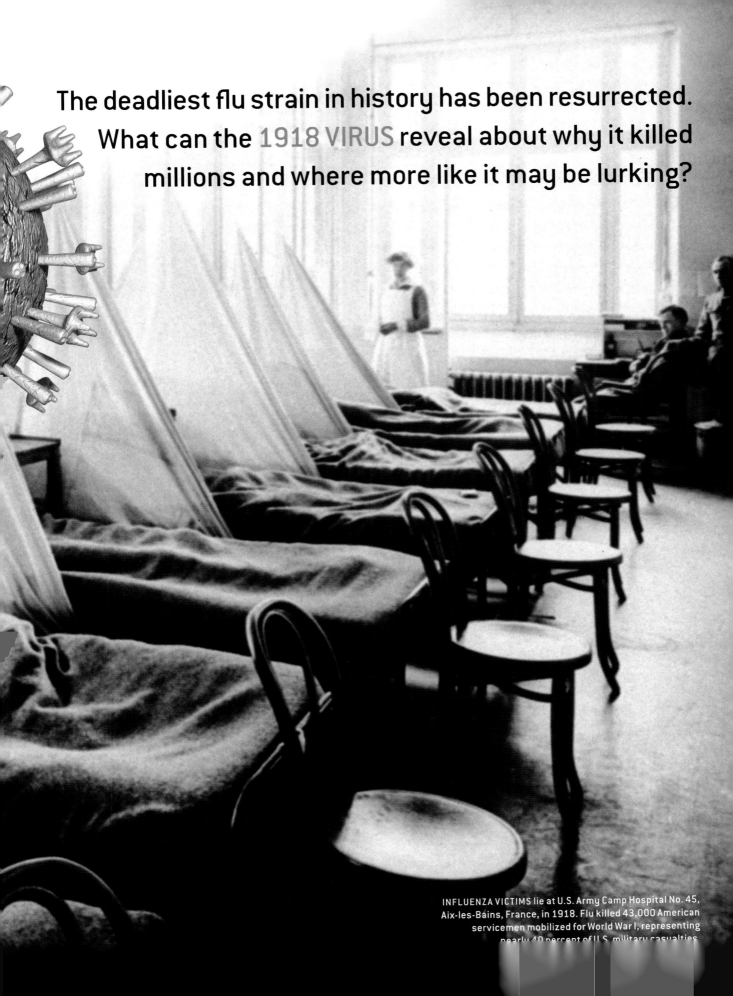

The deadliest flu strain in history has been resurrected. What can the 1918 VIRUS reveal about why it killed millions and where more like it may be lurking?

INFLUENZA VICTIMS lie at U.S. Army Camp Hospital No. 45, Aix-les-Bains, France, in 1918. Flu killed 43,000 American servicemen mobilized for World War I, representing nearly 40 percent of U.S. military casualties.

On September 7, 1918, at the height of World War I, a soldier at an army training camp outside Boston came to sick call with a high fever. Doctors diagnosed him with meningitis but changed their minds the next day when a dozen more soldiers were hospitalized with respiratory symptoms. Thirty-six new cases of this unknown illness appeared on the 16th. Incredibly, by September 23rd, 12,604 cases had been reported in the camp of 45,000 soldiers. By the end of the outbreak, one third of the camp's population would come down with this severe disease, and nearly 800 of them would die. The soldiers who perished often developed a bluish skin color and struggled horribly before succumbing to death by suffocation. Many died less than 48 hours after their symptoms appeared, and at autopsy their lungs were filled with fluid or blood.

Because this unusual suite of symptoms did not fit any known malady, a distinguished pathologist of the era, William Henry Welch, speculated that "this must be some new kind of infection or plague." Yet the disease was neither plague nor even new. It was just influenza. Still, this particularly virulent and infectious strain of the flu virus is thought to have killed as many as 40 million people around the world between 1918 and 1919.

This most lethal flu outbreak in modern history disappeared almost as quickly as it emerged, and its cause was long believed lost to time. No one had preserved samples of the pathogen for later

RED CROSS NURSES in St. Louis carry a flu patient in 1918. Health workers, police and a panicked public donned face masks for protection as the virus swept the country. Nearly a third of all Americans were infected during the pandemic, and 675,000 of them died.

study because influenza would not be identified as a virus until the 1930s. But thanks to incredible foresight by the U.S. Army Medical Museum, the persistence of a pathologist named Johan Hultin, and advances in genetic analysis of old tissue samples, we have been able to retrieve parts of the 1918 virus and study their features. Now, more than 80 years after the horrible natural disaster of 1918–1919, tissues recovered from a handful of victims are answering fundamental questions both about the nature of this pandemic strain and about the workings of influenza viruses in general.

The effort is not motivated merely by historical curiosity. Because influenza viruses continually evolve, new influenza strains continually threaten human populations. Pandemic human flu viruses have emerged twice since 1918—in 1957 and 1968. And flu strains that usually infect only animals have also periodically caused disease in humans, as seen in the recent outbreak of avian influenza in Asia. Our two principal goals are determining what made the 1918 influenza so virulent, to guide development of influenza treatments and preventive measures, and establishing the origin of the pandemic virus, to better target possible sources of future pandemic strains.

Hunting the 1918 Virus

IN MANY RESPECTS, the 1918 influenza pandemic was similar to others before it and since. Whenever a new flu strain emerges with features that have never been encountered by most people's immune systems, widespread flu outbreaks are likely. But certain unique characteristics of the 1918 pandemic have long remained enigmatic.

For instance, it was exceptional in both its breadth and depth. Outbreaks swept across Europe and North America, spreading as far as the Alaskan wilderness and the most remote islands of the Pacific. Ultimately, one third of the world's population may have been infected. The disease was also unusually severe, with death rates of 2.5 to 5 percent—up to 50 times the mortality seen in other influenza outbreaks.

Overview/*The Mystery of 1918*

- The flu pandemic that swept the globe in 1918–1919 was exceptional for the sheer numbers it killed, especially the number of young people who succumbed to the unusually virulent flu virus.
- What made the strain so deadly was a longstanding medical mystery until the authors devised techniques that allowed them to retrieve the 1918 virus's genes from victims' preserved tissues.
- Analysis of those genes and the proteins they encode revealed viral features that could have both suppressed immune defenses and provoked a violent immune reaction in victims, contributing to the high mortality.
- Known bird and mammal influenza hosts are unlikely sources of the pandemic virus, so its origin remains unsolved.

By the fall of 1918 everyone in Europe was calling the disease the "Spanish" influenza, probably because neutral Spain did not impose the wartime censorship of news about the outbreak prevalent in combatant countries. The name stuck, although the first outbreaks, or spring wave, of the pandemic seemingly arose in and around military camps in the U.S. in March 1918. The second, main wave of the global pandemic occurred from September to November 1918, and in many places yet another severe wave of influenza hit in early 1919.

Antibiotics had yet to be discovered, and most of the people who died during the pandemic succumbed to pneumonia caused by opportunistic bacteria that infected those already weakened by the flu. But a subset of influenza victims died just days after the onset of their symp-

the 1918 virus preserved in the victims' lungs. Unfortunately, all attempts to culture live influenza virus from these specimens were unsuccessful.

In 1995 our group initiated an attempt to find the 1918 virus using a different source of tissue: archival autopsy specimens stored at the Armed Forces Institute of Pathology (AFIP). For several years, we had been developing expertise in extracting fragile viral genetic material from damaged or decayed tissue for diagnostic purposes. In 1994, for instance, we were able to use our new techniques to help an AFIP marine mammal pathologist investigate a mass dolphin die-off that had been blamed on red tide. Although the available dolphin tissue samples were badly decayed, we extracted enough pieces of RNA from them to identify a new virus, similar to the one that

damage characteristic of patients who died rapidly. Because the influenza virus normally clears the lungs just days after infection, we had the greatest chance of finding virus remnants in these victims.

The standard practice of the era was to preserve autopsy specimens in formaldehyde and then embed them in paraffin, so fishing out tiny genetic fragments of the virus from these 80-year-old "fixed" tissues pushed the very limits of the techniques we had developed. After an agonizing year of negative results, we found the first influenza-positive sample in 1996, a lung specimen from a soldier who died in September 1918 at Fort Jackson, S.C. We were able to determine the sequence of nucleotides in small fragments of five influenza genes from this sample.

But to confirm that the sequences belonged to the lethal 1918 virus, we kept

After an AGONIZING YEAR of negative results, we found THE FIRST CASE in 1996.

toms from a more severe viral pneumonia—caused by the flu itself—that left their lungs either massively hemorrhaged or filled with fluid. Furthermore, most deaths occurred among young adults between 15 and 35 years old, a group that rarely dies from influenza. Strikingly, people younger than 65 years accounted for more than 99 percent of all "excess" influenza deaths (those above normal annual averages) in 1918–1919.

Efforts to understand the cause of the 1918 pandemic and its unusual features began almost as soon as it was over, but the culprit virus itself remained hidden for nearly eight decades. In 1951 scientists from the University of Iowa, including a graduate student recently arrived from Sweden named Johan Hultin, went as far as the Seward Peninsula of Alaska seeking the 1918 strain [*see box on page 37*]. In November 1918 flu spread through an Inuit fishing village now called Brevig Mission in five days, killing 72 people—about 85 percent of the adult population. Their bodies had since been buried in permafrost, and the 1951 expedition members hoped to find

causes canine distemper, which proved to be the real cause of the dolphin deaths. Soon we began to wonder if there were any older medical mysteries we might solve with our institute's resources.

A descendant of the U.S. Army Medical Museum founded in 1862, the AFIP has grown along with the medical specialty of pathology and now has a collection of three million specimens. When we realized that these included autopsy samples from 1918 flu victims, we decided to go after the pandemic virus. Our initial study examined 78 tissue samples from victims of the deadly fall wave of 1918, focusing on those with the severe lung

looking for more positive cases and identified another one in 1997. This soldier also died in September 1918, at Camp Upton, N.Y. Having a second sample allowed us to confirm the gene sequences we had, but the tiny quantity of tissue remaining from these autopsies made us worry that we would never be able to generate a complete virus sequence.

A solution to our problem came from an unexpected source in 1997: Johan Hultin, by then a 73-year-old retired pathologist, had read about our initial results. He offered to return to Brevig Mission to try another exhumation of 1918 flu victims interred in permafrost. Forty-

THE AUTHORS

JEFFERY K. TAUBENBERGER, ANN H. REID and THOMAS G. FANNING work together at the Armed Forces Institute of Pathology in Rockville, Md. In 1993 Taubenberger, a molecular pathologist, helped to create a laboratory there devoted to molecular diagnostics—identifying diseases by their genetic signatures rather than by the microscopic appearance of patients' tissue samples. Early work by Reid, a molecular biologist, led the group to devise the techniques for extracting DNA and RNA from damaged or decayed tissue that allowed them to retrieve bits and pieces of 1918 flu virus genes from archived autopsy specimens. Fanning, a geneticist with expertise in the evolution of genomes, helped to analyze the genes' relationships to other animal and human flu viruses. The authors wish to note that the opinions expressed in this article are their own and do not represent the views of the Department of Defense or the AFIP.

FLU HIJACKS HOSTS TO REPLICATE AND EVOLVE

Influenza is a small and simple virus—just a hollow lipid ball studded with a few proteins and bearing only eight gene segments (*below*). But that is all it needs to induce the cells of living hosts to make more viruses (*bottom*). One especially important protein on influenza's surface, hemagglutinin (HA), allows the virus to enter cells. Its shape determines which hosts a flu virus strain can infect. Another protein, neuraminidase (NA), cuts newly formed viruses loose from an infected cell, influencing how efficiently the virus can spread. Slight changes in these and other flu proteins can help the virus infect new kinds of hosts and evade immune attack. The alterations can arise through mistakes that occur while viral genes are being copied. Or they can be acquired in trade when the genes of two different flu viruses infecting the same cell intermingle (*right*).

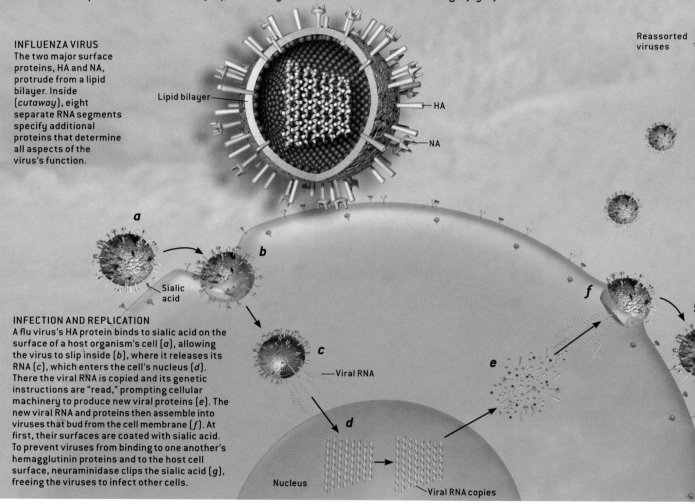

INFLUENZA VIRUS
The two major surface proteins, HA and NA, protrude from a lipid bilayer. Inside (*cutaway*), eight separate RNA segments specify additional proteins that determine all aspects of the virus's function.

Lipid bilayer

HA

NA

Reassorted viruses

a

Sialic acid

b

f

INFECTION AND REPLICATION
A flu virus's HA protein binds to sialic acid on the surface of a host organism's cell (*a*), allowing the virus to slip inside (*b*), where it releases its RNA (*c*), which enters the cell's nucleus (*d*). There the viral RNA is copied and its genetic instructions are "read," prompting cellular machinery to produce new viral proteins (*e*). The new viral RNA and proteins then assemble into viruses that bud from the cell membrane (*f*). At first, their surfaces are coated with sialic acid. To prevent viruses from binding to one another's hemagglutinin proteins and to the host cell surface, neuraminidase clips the sialic acid (*g*), freeing the viruses to infect other cells.

c

Viral RNA

e

d

Nucleus

Viral RNA copies

six years after his first attempt, with permission from the Brevig Mission Council, he obtained frozen lung biopsies of four flu victims. In one of these samples, from a woman of unknown age, we found influenza RNA that provided the key to sequencing the entire genome of the 1918 virus.

More recently, our group, in collaboration with British colleagues, has also been surveying autopsy tissue samples from 1918 influenza victims from the Royal London Hospital. We have been able to analyze flu virus genes from two

of these cases and have found that they were nearly identical to the North American samples, confirming the rapid worldwide spread of a uniform virus. But what can the sequences tell us about the virulence and origin of the 1918 strain? Answering those questions requires a bit of background about how influenza viruses function and cause disease in different hosts.

Flu's Changing Face
EACH OF THE THREE novel influenza strains that caused pandemics in the past

100 years belonged to the type A group of flu viruses. Flu comes in three main forms, designated A, B and C. The latter two infect only humans and have never caused pandemics. Type A influenza viruses, on the other hand, have been found to infect a wide variety of animals, including poultry, swine, horses, humans and other mammals. Aquatic birds, such as ducks, serve as the natural "reservoir" for all the known subtypes of influenza A, meaning that the virus infects the bird's gut without causing symptoms. But these wild avian strains

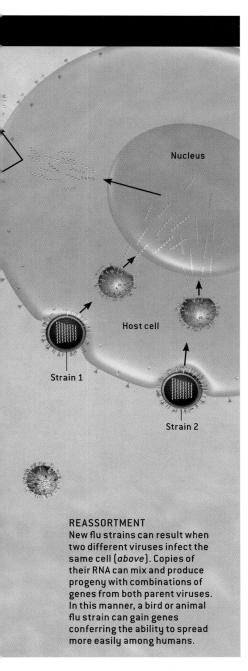

REASSORTMENT
New flu strains can result when two different viruses infect the same cell (*above*). Copies of their RNA can mix and produce progeny with combinations of genes from both parent viruses. In this manner, a bird or animal flu strain can gain genes conferring the ability to spread more easily among humans.

lular machinery, inducing it to manufacture new viral proteins and additional copies of viral RNA. These pieces then assemble themselves into new viruses that escape the host cell, proceeding to infect other cells. No proofreading mechanism ensures that the RNA copies are accurate, so mistakes leading to new mutations are common. What is more, should two different influenza virus strains infect the same cell, their RNA segments can mix freely there, producing progeny viruses that contain a combination of genes from both the original viruses. This "reassortment" of viral genes is an important mechanism for generating diverse new strains.

Different circulating influenza A viruses are identified by referring to two signature proteins on their surfaces. One is hemagglutinin (HA), which has at least 15 known variants, or subtypes. Another is neuraminidase (NA), which has nine subtypes. Exposure to these proteins produces distinctive antibodies in a host, thus the 1918 strain was the first to be named, "H1N1," based on antibodies found in the bloodstream of pandemic survivors. Indeed, less virulent descendants of H1N1 were the predominant circulating flu strains until 1957, when an H2N2 virus emerged, causing a pandemic. Since 1968, the H3N2 subtype, which provoked the pandemic that year, has predominated.

The HA and NA protein subtypes present on a given influenza A virus are more than just identifiers; they are essential for viral reproduction and are primary targets of an infected host's immune system. The HA molecule initiates infection by binding to receptors on the surface of certain host cells. These tend to be respiratory lining cells in mammals and intestinal lining cells in birds. The NA protein enables new virus copies to escape the host cell so they can go on to infect other cells.

After a host's first exposure to an HA subtype, antibodies will block receptor binding in the future and are thus very effective at preventing reinfection with the same strain. Yet flu viruses with HA subtypes that are new to humans periodically appear, most likely through reassortment with the extensive pool of influenza viruses infecting wild birds. Normally, influenza HAs that are adapted to avian hosts bind poorly to the cell-surface receptors prevalent in the human respiratory tract, so an avian virus's HA binding affinity must be somewhat modified before the virus can replicate and spread efficiently in humans. Until recently, existing evidence suggested that a wholly avian influenza virus probably could not directly infect humans, but 18 people were infected with an avian H5N1 influenza virus in Hong Kong in 1997, and six died.

Outbreaks of an even more pathogenic version of that H5N1 strain became widespread in Asian poultry in 2003 and 2004, and more than 30 people infected with this virus have died in Vietnam and Thailand.

The virulence of an influenza virus once it infects a host is determined by a complex set of factors, including how readily the virus enters different tissues, how quickly it replicates, and the violence of the host's immune response to the intruder. Thus, understanding exactly what made the 1918 pandemic influenza strain so infectious and so virulent could yield great insight into what makes any influenza strain more or less of a threat.

A Killer's Face
WITH THE 1918 RNA we have retrieved, we have used the virus's own genes as recipes for manufacturing its component parts—essentially re-creating pieces of the killer virus itself. The first of these we were eager to examine was the hemagglutinin protein, to look for features that might explain the exceptional virulence of the 1918 strain.

We could see, for example, that the part of the 1918 HA that binds with a host cell is nearly identical to the binding site of a wholly avian influenza HA [*see illustration on page 35*]. In two of the 1918 isolates, this receptor-binding site differs from an avian form by only one amino acid building block. In the other three isolates, a second amino acid is also altered. These seemingly subtle mutations may represent the min-

can mutate over time or exchange genetic material with other influenza strains, producing novel viruses that are able to spread among mammals and domestic poultry.

The life cycle and genomic structure of influenza A virus allow it to evolve and exchange genes easily. The virus's genetic material consists of eight separate RNA segments encased in a lipid membrane studded with proteins [*see top illustration on opposite page*]. To reproduce, the virus binds to and then enters a living cell, where it commandeers cel-

When analyzing the genes of the 1918 virus revealed no definitive reasons for the pandemic strain's virulence, our group turned to reverse genetics—a method of understanding the function of genes by studying the proteins they encode. In collaboration with scientists from the Mount Sinai School of Medicine, the Centers for Disease Control and Prevention, the U.S. Department of Agriculture, the University of Washington and the Scripps Research Institute, we "built" influenza viruses containing one or more of the 1918 virus's genes, so we could see how these recombinant viruses behaved in animals and human cell cultures.

To construct these viruses, we employed a new technique called plasmid-based reverse genetics, which requires first making DNA copies of flu genes that normally exist in RNA form. Each DNA gene copy is then inserted into a tiny ring of DNA called a plasmid. Different combinations of these plasmids can be injected into living cells, where cellular machinery will execute the genetic instructions they bear and manufacture flu viruses with only the desired combination of genes.

Reverse genetics not only allows us to study the 1918 virus, it will allow scientists in the U.S. and Europe to explore how great a threat the H5N1 avian flu virus poses to humans. Since January 2004, that strain—which is now present in birds in 10 Asian countries—has infected more than 40 people, killing more than 30 of them. One of the casualties was a mother who is believed to have contracted the virus from her daughter, rather than directly from a bird.

Such human-to-human transmission could suggest that in their case the avian virus had adapted to be more easily

PLASMID-BASED reverse genetics lets scientists custom-manufacture flu viruses. DNA copies of genes from two different flu strains (*blue* and *red*) are inserted into DNA rings called plasmids. The gene-bearing plasmids are then injected into a culture of living cells, which manufacture whole flu viruses containing the desired combination of genes.

spread between humans, either by mutating or by acquiring new genes through reassortment with a circulating human flu strain. That dreaded development would increase the possibility of a human pandemic. Hoping to predict and thereby prevent such a disaster, scientists at the CDC and Erasmus University in the Netherlands are planning to test combinations of H5N1 with current human flu strains to assess the likelihood of their occurring naturally and their virulence in people.

What these experiments will reveal, as in our group's work with the 1918 virus genes, is crucial to understanding how influenza pandemics form and why they cause disease. Some observers have questioned the safety of experimenting with lethal flu strains, but all of this research is conducted in secure laboratories designed specifically to deal with highly pathogenic influenza viruses.

What is more, re-creating the 1918 virus proteins enabled us to establish that currently available antiviral drugs, such as amantadine or the newer neuraminidase inhibitors, such as oseltamivir (Tamiflu), would be effective against the 1918 strain in the case of an accidental infection. The H5N1 viruses are also sensitive to the neuraminidase inhibitors.

Scientists in the U.S. and U.K. also recently employed plasmid-based reverse genetics to create a seed strain for a human vaccine against H5N1. They made a version of the H5N1 virus lacking the wild strain's most deadly features, so that manufacturers could safely use it to produce a vaccine [see "The Scientific American 50," December 2004]. Clinical trials of that H5N1 vaccine were scheduled to begin at the end of 2004.

—*J.K.T., A.H.R. and T.G.F.*

imal change necessary to allow an avian-type HA to bind to mammalian-type receptors.

But while gaining a new binding affinity is a critical step that allows a virus to infect a new type of host, it does not necessarily explain why the 1918 strain was so lethal. We turned to the gene sequences themselves, looking for features that could be directly related to virulence, including two known mutations in other flu viruses. One involves the HA gene: to become active in a cell, the HA

protein must be cleaved into two pieces by a gut-specific protein-cutting enzyme, or protease, supplied by the host. Some avian H5 and H7 subtype viruses acquire a gene mutation that adds one or more basic amino acids to the cleavage site, allowing HA to be activated by ubiquitous proteases. In chickens and other birds, infection by such a virus causes disease in multiple organs and even the central nervous system, with a very high mortality rate. This mutation has been observed in the H5N1 viruses currently

circulating in Asia. We did not, however, find it in the 1918 virus.

The other mutation with a significant effect on virulence has been seen in the NA gene of two influenza virus strains that infect mice. Again, mutations at a single amino acid appear to allow the virus to replicate in many different body tissues, and these flu strains are typically lethal in laboratory mice. But we did not see this mutation in the NA of the 1918 virus either.

Because analysis of the 1918 virus's

genes was not revealing any characteristics that would explain its extreme virulence, we initiated a collaborative effort with several other institutions to re-create parts of the 1918 virus itself so we could observe their effects in living tissues.

A new technique called plasmid-based reverse genetics allows us to copy 1918 viral genes and then combine them with the genes of an existing influenza strain, producing a hybrid virus. Thus, we can take an influenza strain adapted to mice, for example, and give it different combinations of 1918 viral genes. Then, by infecting a live animal or a human tissue culture with this engineered virus, we can see which components of

a tissue culture of human lung cells, we found that a virus with the 1918 NS1 gene was indeed more effective at blocking the host's type I IFN system.

To date, we have produced recombinant influenza viruses containing between one and five of the 1918 genes. Interestingly, we found that any of the recombinant viruses possessing both the 1918 HA and NA genes were lethal in mice, causing severe lung damage similar to that seen in some of the pandemic fatalities. When we analyzed these lung tissues, we found signatures of gene activation involved in common inflammatory responses. But we also found higher than normal activation of

unclear, this protein may have played a key role in the 1918 strain's virulence.

These ongoing experiments are providing a window to the past, helping scientists understand the unusual characteristics of the 1918 pandemic. Similarly, these techniques will be used to study what types of changes to the current H5N1 avian influenza strain might give that extremely lethal virus the potential to become pandemic in humans [*see box on opposite page*]. An equally compelling question is how such virulent strains emerge in the first place, so our group has also been analyzing the 1918 virus's genes for clues about where it might have originated.

Seemingly subtle mutations may allow an AVIAN hemagglutinin to bind to MAMMALIAN receptors.

the pandemic strain might have been key to its pathogenicity.

For instance, the 1918 virus's distinctive ability to produce rapid and extensive damage to both upper and lower respiratory tissues suggests that it replicated to high numbers and spread quickly from cell to cell. The viral protein NS1 is known to prevent production of type I interferon (IFN)—an "early warning" system that cells use to initiate an immune response against a viral infection. When we tested recombinant viruses in

genes associated with the immune system's offensive soldiers, T cells and macrophages, as well as genes related to tissue injury, oxidative damage, and apoptosis, or cell suicide.

More recently, Yoshihiro Kawaoka of the University of Wisconsin–Madison reported similar experiments with 1918 flu genes in mice, with similar results. But when he tested the HA and NA genes separately, he found that only the 1918 HA produced the intensive immune response, suggesting that for reasons as yet

Seeking the Source

THE BEST APPROACH to analyzing the relationships among influenza viruses is phylogenetics, whereby hypothetical family trees are constructed using viral gene sequences and knowledge of how often genes typically mutate. Because the genome of an influenza virus consists of eight discrete RNA segments that can move independently by reassortment, these evolutionary studies must be performed separately for each gene segment.

We have completed analyses of five of the 1918 virus's eight RNA segments, and so far our comparisons of the 1918 flu genes with those of numerous human, swine and avian influenza viruses always place the 1918 virus within the human and swine families, outside the avian virus group [*see box on next page*]. The 1918 viral genes do have some avian features, however, so it is probable that the virus originally emerged from an avian reservoir sometime before 1918. Clearly by 1918, though, the virus had acquired enough adaptations to mammals to function as a human pandemic virus. The question is, where?

When we analyzed the 1918 hemagglutinin gene, we found that the sequence has many more differences from

| Human-adapted | Avian-adapted | 1918 Flu |
| H3 | H5 | H1 |

Amino
acid
change

HEMAGGLUTININ (HA) of the 1918 flu strain was re-created from its gene sequence by the authors' collaborators so they could examine the part that binds to a host cell's sialic acid and allows the virus to enter the cell. HA binding sites usually are shaped differently enough to bar cross-species infection. For instance, the human-adapted H3-type HA has a wide cavity in the middle of its binding site (*left*), whereas the avian H5 cavity (*center*) is narrow. The 1918 H1-type HA (*right*) more closely resembles the avian form, with only a few minor differences in the sequence of its amino acid building blocks. One of these alterations (*above right*) slightly widens the central cavity, apparently just enough to have allowed a flu virus with this avian-type HA to infect hundreds of millions of humans in 1918–1919.

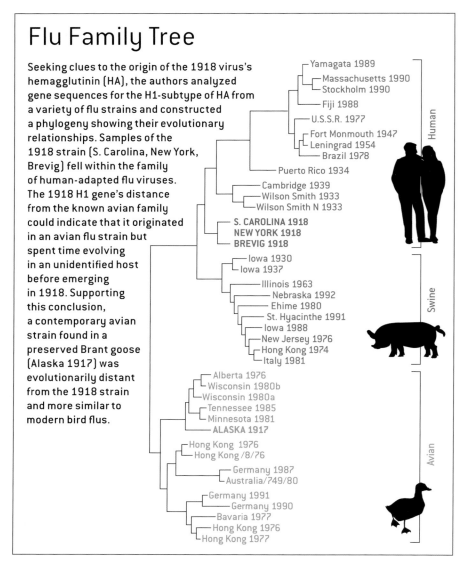

Flu Family Tree

Seeking clues to the origin of the 1918 virus's hemagglutinin (HA), the authors analyzed gene sequences for the H1-subtype of HA from a variety of flu strains and constructed a phylogeny showing their evolutionary relationships. Samples of the 1918 strain (S. Carolina, New York, Brevig) fell within the family of human-adapted flu viruses. The 1918 H1 gene's distance from the known avian family could indicate that it originated in an avian flu strain but spent time evolving in an unidentified host before emerging in 1918. Supporting this conclusion, a contemporary avian strain found in a preserved Brant goose (Alaska 1917) was evolutionarily distant from the 1918 strain and more similar to modern bird flus.

Yamagata 1989
Massachusetts 1990
Stockholm 1990
Fiji 1988
U.S.S.R. 1977
Fort Monmouth 1947
Leningrad 1954
Brazil 1978
Puerto Rico 1934
Cambridge 1939
Wilson Smith 1933
Wilson Smith N 1933
S. CAROLINA 1918
NEW YORK 1918
BREVIG 1918
Iowa 1930
Iowa 1937
Illinois 1963
Nebraska 1992
Ehime 1980
St. Hyacinthe 1991
Iowa 1988
New Jersey 1976
Hong Kong 1974
Italy 1981
Alberta 1976
Wisconsin 1980b
Wisconsin 1980a
Tennessee 1985
Minnesota 1981
ALASKA 1917
Hong Kong 1976
Hong Kong /8/76
Germany 1987
Australia/749/80
Germany 1991
Germany 1990
Bavaria 1977
Hong Kong 1976
Hong Kong 1977

Human

Swine

Avian

avian sequences than do the 1957 H2 and 1968 H3 subtypes. Thus, we concluded, either the 1918 HA gene spent some length of time in an intermediate host where it accumulated many changes from the original avian sequence, or the gene came directly from an avian virus, but one that was markedly different from known avian H1 sequences.

To investigate the latter possibility that avian H1 genes might have changed substantially in the eight decades since the 1918 pandemic, we collaborated with scientists from the Smithsonian Institution's Museum of Natural History and Ohio State University. After examining many preserved birds from the era, our group isolated an avian subtype H1 influenza strain from a Brant goose

collected in 1917 and stored in ethanol in the Smithsonian's bird collections. As it turned out, the 1917 avian H1 sequence was closely related to modern avian North American H1 strains, suggesting that avian H1 sequences have changed little over the past 80 years. Extensive sequencing of additional wild bird H1 strains may yet identify a strain more similar to the 1918 HA, but it may be that no avian H1 will be found resembling the 1918 strain because, in fact, the HA did not reassort directly from a bird strain.

In that case, it must have had some intermediate host. Pigs are a widely suggested possibility because they are known to be susceptible to both human and avian viruses. Indeed, simultaneous outbreaks of influenza were seen in hu-

mans and swine during the 1918 pandemic, but we believe that the direction of transmission was most probably from humans to pigs. There are numerous examples of human influenza A virus strains infecting swine since 1918, but swine influenza strains have been isolated only sporadically from humans. Nevertheless, to explore the possibility that the 1918 HA may have started as an avian form that gradually adapted to mammalian hosts in swine, we looked at a current example of how avian viruses evolve in pigs—an avian H1N1 influenza lineage that has become established in European swine over the past 25 years. We found that even 20 years of evolution in swine has not resulted in the number of changes from avian sequences exhibited by the 1918 pandemic strain.

When we applied these types of analyses to four other 1918 virus genes, we came to the same conclusion: the virus that sparked the 1918 pandemic could well have been an avian strain that was evolutionarily isolated from the typical wild waterfowl influenza gene pool for some time—one that, like the SARS coronavirus, emerged into circulation among humans from an as yet unknown animal host.

Future Investigations

OUR ANALYSES of five RNA segments from the 1918 virus have shed some light on its origin and strongly suggest that the pandemic virus was the common ancestor of both subsequent human and swine H1N1 lineages, rather than having emerged from swine. To date, analyzing the viral genes has offered no definitive clue to the exceptional virulence of the 1918 virus strain. But experiments with engineered viruses containing 1918 genes indicate that certain of the 1918 viral proteins could promote rapid virus replication and provoke an intensely destructive host immune response.

In future work, we hope that the 1918 pandemic virus strain can be placed in the context of influenza viruses that immediately preceded and followed it. The direct precursor of the pandemic virus, the first or spring wave virus strain, lacked the autumn wave's exceptional virulence and seemed to spread less easily. At present, we are seeking influenza RNA samples from victims of the spring wave to identify any genetic differences between the two strains that might help elucidate why the autumn wave was more severe. Similarly, finding pre-1918 human influenza RNA samples would clarify which gene segments in the 1918 virus were completely novel to humans. The unusual mortality among young people during the 1918 pandemic might be explained if the virus shared features with earlier circulating strains to which older people had some immunity. And finding samples of H1N1 from the 1920s and later would help us understand the 1918 virus's subsequent evolution into less virulent forms.

We must remember that the mechanisms by which pandemic flu strains originate are not yet fully understood. Because the 1957 and 1968 pandemic strains had avian-like HA proteins, it seems most likely that they originated in the direct reassortment of avian and human virus strains. The actual circumstances of those reassortment events have never been identified, however, so no one knows how long it took for the novel strains to develop into human pandemics.

The 1918 pandemic strain is even more puzzling, because its gene sequences are consistent neither with direct reassortment from a known avian strain nor with adaptation of an avian strain in swine. If the 1918 virus should prove to have acquired novel genes through a different mechanism than subsequent pandemic strains, this could have important public health implications. An alternative origin might even have contributed to the 1918 strain's exceptional virulence. Sequencing of many more avian influenza viruses and research into alternative intermediate hosts other than swine, such as poultry, wild birds or horses, may provide more clues to the 1918 pandemic's source. Until the origins of such strains are better understood, detection and prevention efforts may overlook the beginning of the next pandemic. ⌐SA⌐

Persistence Pays Off

Visiting Alaska in the summer of 1949, Swedish medical student Johan Hultin met Lutheran missionaries in Fairbanks who told him of the 1918 flu pandemic's toll on Inuit villages. One, a tiny settlement on the Seward Peninsula called Teller Mission, was all but wiped out in November 1918. Overwhelmed missionaries had to call in the U.S. Army to help bury 72 victims' bodies in a mass grave, which they marked by two crosses.

Haunted by the story, Hultin (right, center and below) headed to the University of Iowa to begin his doctoral studies in microbiology. There he kept thinking about the 1918 pandemic and wondering if the deadly virus that caused it could be retrieved for study from bodies that may have been preserved by the Alaskan permafrost. In the summer of 1951, Hultin convinced

HULTIN in Brevig grave, 1951

two Iowa faculty, a virologist and a pathologist, to visit the village, then called Brevig Mission. With permission from tribal elders, the scientists excavated the grave and obtained tissue specimens from what remained of several victims' lungs.

Back in Iowa, the team tried and tried to grow live virus from the specimens but never could. In retrospect, that was perhaps just as well since biological containment equipment for dangerous pathogens did not exist at the time.

Hultin's disappointment led him to abandon his Ph.D. and become a pathologist instead. Retired and living in San Francisco in 1997, Hultin read our group's first published description of the 1918 genes we retrieved from autopsy specimens, and it rekindled his hope of finding the entire 1918 virus. He wrote to me, eager to try to procure new lung specimens from Brevig Mission for us to work with. He offered to leave immediately for Alaska, and I agreed.

At the same time, Hultin tracked down his 1951 expedition mates to ask if they had kept any of the original Brevig specimens. We reasoned that those tissue samples obtained just 33 years after the pandemic and then preserved might be in better condition than specimens taken later. As it turned out, one of Hultin's colleagues had kept the material in storage for years but finally deemed it useless and threw it out. He had disposed of the last specimens just the year before, in 1996.

Fortunately, Hultin once again got permission from the Brevig Mission Council to excavate the 1918 grave in August 1997. And this time he found the body of a young woman who had been obese in life. Hultin said later that he knew instantly her tissue samples would contain the 1918 virus—together with the cold temperature, her thick layer of fat had almost perfectly preserved her lungs. He was right, and her tissue provided us with the entire genome of the 1918 pandemic virus. —J.K.T.

HULTIN in Brevig grave, 1997

MORE TO EXPLORE

Devil's Flu: The World's Deadliest Influenza Epidemic and the Scientific Hunt for the Virus That Caused It. Pete Davies. Henry Holt and Co., 2000.

America's Forgotten Pandemic: The Influenza of 1918. Second edition. Alfred W. Crosby. Cambridge University Press, 2003.

The Origin of the 1918 Pandemic Influenza Virus: A Continuing Enigma. Ann H. Reid and Jeffery K. Taubenberger in Journal of General Virology, Vol. 84, Part 9, pages 2285–2292; September 2003.

Global Host Immune Response: Pathogenesis and Transcriptional Profiling of Type A Influenza Viruses Expressing the Hemagglutinin and Neuraminidase Genes from the 1918 Pandemic Virus. J. C. Kash, C. F. Basler, A. Garcia-Sastre, V. Carter, R. Billharz, D. E. Swayne, R. M. Przygodzki, J. K. Taubenberger, M. G. Katze and T. M. Tumpey in Journal of Virology, Vol. 78, No. 17, pages 9499–9511; September 2004.

Questions for Review

"Capturing a Killer Flu Virus"

By Jeffery K. Taubenberger, Ann H. Reid, and Thomas G. Fanning

TESTING YOUR COMPREHENSION

1. When did influenza pandemics occur in the 20th century?
   a. 1918
   b. 1918, 1957, 1968
   c. 1918, 1980
   d. 1918-19

2. The 1918 influenza virus originated in
   a. Alaska.
   b. the continental United States.
   c. pigs.
   d. Spain.

3. Hemagglutinin (HA)
   a. attaches viruses to a host cell.
   b. releases viruses from a host cell.
   c. replicates viral genes.
   d. recombines genes from different viruses.

4. Influenza virus has
   a. a single strand of RNA.
   b. a double-stranded molecule of RNA.
   c. four stands of RNA.
   d. eight strands of RNA.

5. If two different influenza viruses infect a cell,
   a. their RNA segments can mix, producing a virus with a mixture of genes.
   b. both viruses die.
   c. one virus reproduces and inhibits the other virus.
   d. no one knows what will happen.

6. The current predominant strain of influenza is
   a. H1N1.
   b. H2N2.
   c. H3N2.
   d. H5N1.

7. One reason for the 1918 virus's virulence is it
   a. inhibits the host's antibody formation.
   b. lyses host cells.
   c. blocks host's interferon.
   d. causes pneumonia.

8. The genome from the 1918 influenza virus was isolated from a(an)
   a. Brandt's goose.
   b. Inuit woman.
   c. pig.
   d. South Carolina soldier.

9. What group of influenza viruses has caused all the pandemics in the last 100 years?
   a. A
   b. B
   c. C
   d. H1N1

10. Neuraminidase (NA)
    a. attaches viruses to a host cell.
    b. removes sialic acid from the newly released virus.
    c. replicates virus genes.
    d. recombines genes from different viruses.

MICROBIOLOGY IN SOCIETY

1. Of what value is studying a virus from 1918?

2. What caused most of the deaths in the 1918-19 influenza pandemic? Could this happen again?

3. Pathogenic viruses are being made by reverse genetics. Should genetic modification (reverse genetics) of viruses be allowed?

THINKING ABOUT MICROBIOLOGY

1. The influenza family tree shown on page 36 is called a cladogram. How is a cladogram made? Which virus is more closely related to Puerto Rico 1934: Cambridge 1939 or Fiji 1988?

2. How do new strains of influenza virus evolve?

3. Explain how type A influenza viruses are named.

WRITING ABOUT MICROBIOLOGY

1. In 2003 and 2004, outbreaks of influenza occurred in Asian poultry and more than 30 people died from this avian influenza (H5N1) virus. This virus was not transmitted between humans. Using the 1918-virus model, what would be required to have a human pandemic?

2. The 1918 influenza pandemic was unusual because so many deaths occurred among 15–35 year olds. Explain why this could occur.

3. What is reverse genetics?

Answers can be found on The Microbiology Place website. Go to www.microbiologyplace.com, click on the cover of your textbook, and type in your login name and password (using the access code found in the front pages of your textbook). Then, click on Current Issues Magazine Answers on the left navigation bar.

Can Chlamydia Be Stopped?

By David M. Ojcius, Toni Darville and Patrik M. Bavoil

Chlamydia is a rampant sexually transmitted disease, the world's leading cause of preventable blindness and a possible contributor to heart disease. Recent discoveries are suggesting new ways to curtail its spread

Ask the average American about chlamydia, and you will probably evoke an uneasy cringe. Most people think immediately of one of the world's most common sexually transmitted infections (STIs). But the term actually refers to two genera of tiny bacteria that can ignite a variety of serious illnesses.

Ask a poor mother in Africa about chlamydia, and she may tell you that flies transmitting this infection gave her two young children the painful eye condition known as conjunctivitis. This illness—caused by a strain of *Chlamydia trachomatis* (the species that also causes STIs)—can lead to trachoma, a potentially blinding disease. In industrial countries, an airborne species, *Chlamydophila pneumoniae*, causes colds, bronchitis and about 10 percent of pneumonias acquired outside of hospitals. Researchers have even drawn tentative links between *C. pneumoniae* and atherosclerosis, the artery-narrowing condition that leads to heart attacks and strokes.

Because chlamydiae are bacteria, antibiotics can thwart the infections they produce. Unfortunately, the illnesses often go undetected and untreated, for various reasons. The genital infections rarely produce symptoms early on. And in developing countries where trachoma is a concern, people often lack access to adequate treatment and hygiene. As a result, many of the estimated 600 million people infected with one or more chlamydial strains will go without medical care until the consequences have become irreversible.

It is unrealistic to expect that doctors will ever identify all individuals who have the STI or that improved hygiene will soon wipe out the trachoma-causing bacteria in developing countries. For these reasons, the best hope for curtailing the spread of these ailments is to develop an effective vaccine or other preventive treatments. To discover agents able to block infections before they start, scientists need to know more about how chlamydiae replicate, incite disease and function at a molecular level. But that information has been hard to come by. These bugs are wily. Not only do they have varied strategies for evading the body's immune system, they also are notori-

SEVERAL PARTS of the body, including the eyes, lungs and reproductive tract, are vulnerable to chlamydial infection.

ously difficult to study in the laboratory. In the past five years, however, new research—including the complete sequencing of the genomes of several chlamydial strains—has helped scientists begin to address these obstacles. The resulting discoveries are renewing hope for developing new prevention strategies.

Silent Injury

ONE MAJOR IMPEDIMENT to the production of a vaccine is chlamydia's surreptitious way of wreaking havoc on the body. The microbes that cause tetanus or cholera swamp tissues with toxins that damage or kill vulnerable cells. Chlamydiae, in contrast, do not damage tissues directly. Rather they elicit an enthusiastic immune response that attempts to rein in the infection through inflammation for as long as the bacteria remain in the body—even at low levels. Ironically, this way of fighting the infection actually brings on the long-term damage. Vaccines prevent illness by priming the immune system to react strongly to specific disease-causing agents, but in this case, the inflammatory component of such a response could do more harm than good.

Whether in the genital tract, eyelids or elsewhere, inflammation begins when certain cells of the host immune system secrete factors called cytokines—small signaling proteins that attract additional defensive cells to the site of infection. The attracted cells and the cytokines try to wall off the area to prevent the bacteria's spread. In the skin, this process gives rise to familiar outward manifestations of inflammation: redness, swelling and heat. At the same time, the inflammatory cytokines help to trigger the tissue repair response called fibrosis, which can lead to scarring.

In the genital tract, the early inflammation is not obvious. Of the 3.5 million Americans infected with sexually transmitted chlamydia every year, 85 to 90 percent show no symptoms. Men, whose inflammation occurs in the penis, may experience slight pain during urination; women may feel nothing as the bacteria move up the genital tract into the fallopian tubes. Unaware of the problem, these individuals inadvertently pass the bugs along. Indeed, a woman may not learn of her infection until she tries to become pregnant and realizes she is infertile. In other cases, persistent inflammation and scarring of the fallopian tubes causes chronic pelvic pain or increases the chances of ectopic, or tubal, pregnancy—the leading cause of first-trimester pregnancy-related deaths in the U.S.

Inflammation of the eyelids is more immediately obvious. Such infections afflict an estimated 150 million people living in developing countries with hot climates; there treatments may be scarce, and flies and gnats can readily transmit the bacteria between people's infrequently washed hands and faces. (Trachoma does not occur in the U.S. or western Europe because of better public health systems.) When infections scar the inside of the upper eyelid repeatedly over many years, the eyelid may begin to turn under, pointing the eyelashes inward where they can scratch the cornea. Unchecked, the corneal damage can cause blindness decades after the initial infection.

Given that inflammation accounts for most of chlamydia's ill effects, those who are striving to develop a vaccine must find a way to control the bacteria without inducing a strong inflammatory reaction. Ideally, any intervention would fine-tune the inflammatory response—evoking it just enough to help the body's other immune defenses eliminate the bacteria.

Much research on infections caused by chlamydia and other pathogens is focusing on factors that either initiate secretion of the inflammatory cytokines or dampen the inflammatory response once the infection has been cleared. Over the past few years, investigators have discovered small molecules that normally stimulate or inhibit these responses in the body. The next step will be to develop compounds that are able to regulate the activities of these molecules. These agents might be delivered to shut down inflammation artificially after an antibiotic has been administered to control the bacteria.

Hanging Around

BEYOND INDUCING INFLAMMATION, chlamydiae have other properties that impede development of an effective vaccine. For instance, once you get mumps or measles—or the vaccines against them—you are immune for life. Not so with chlamydia. The body has a hard time eliminating the bacteria completely, and natural immunity after a bout with the microbes lasts only about six months. Hence, an infection that has apparently disappeared may flare up again months or years later, and little protection remains against new outbreaks. If the body's natural response to infection cannot confer long-term protection, it seems likely that a vaccine that merely mimicked this response would fail as well. To be successful, a vaccine would have to elicit defenses that were more powerful than those occurring naturally without triggering excessive inflammation.

One way that vaccines or natural immune responses to an initial infection protect against future colonization by certain microorganisms is by inducing the body to produce so-called memory B lymphocytes targeted to those specific invaders. These immune cells patrol the body throughout its lifetime, ready to secrete antibody molecules that can in turn latch onto any new bugs and mark them for destruction before they in-

Overview/*Too Little, Too Late*

- Chlamydia has many modes of attack. Untreated infections have blinded more than two million people worldwide, leave more than 10,000 women in the U.S. sterile every year, and account for 10 percent of pneumonia cases in industrial countries.
- Most people affected by chlamydia are not treated with antibiotics until after the damage is done; either they do not notice their symptoms right away, or they do not have access to adequate hygiene or health care.
- Global sex education campaigns and improved hygiene can certainly help limit the bacterium's spread, but other preventive measures such as vaccines are probably the only way to stamp out the disease entirely.

CHLAMYDIA IS NOT JUST AN STI

SPECIES	DISEASE	DISTRIBUTION	MODE OF TRANSMISSION	NUMBERS AFFECTED
Chlamydophila pneumoniae	Pneumonia; possibly atherosclerosis	Worldwide	Inhalation of the bacterium within aerosols produced when an infected person coughs	Causes about 10 percent of pneumonia cases in developed countries, including up to 300,000 new cases in the U.S. every year
Chlamydophila psittaci	Psittacosis, a flulike infection of the lungs that can cause inflammation of the liver, heart and brain	Worldwide	Inhalation of the bacterium in aerosols or dust; a bite from or handling the plumage or tissues of an infected bird	Common in wild and domestic birds; rare but potentially fatal when transmitted to humans; 50 to 100 new human cases in the U.S. every year
Chlamydia trachomatis (Different strains cause different disorders.)	Trachoma, a painful eye infection that begins as conjunctivitis and leads to scarring of the cornea and possible blindness	Southeast Asia, South America, India, Middle East, Africa; rare in the U.S.	Direct contact with bodily secretions of infected people or contact with carrier flies or clothing contaminated with such secretions	More than 100 million people worldwide have trachoma, and 2 million are blind as a result of it; virtually no incidence in areas with adequate hygiene
C. trachomatis	Sexually transmitted infection (STI) of the adult genital tract; can cause conjunctivitis and pneumonia in newborns	Worldwide	Sexual contact; newborns acquire the bacterium from their infected mothers while passing through the birth canal	50 million to 90 million new STIs occur globally every year; in the U.S. alone, 3.5 million new infections and more than 10,000 cases of female infertility
C. trachomatis	*Lymphogranuloma venereum*, an STI of the lymph glands in the genital area	Asia, Africa, South America, Central America; rare in the U.S.	Sexual contact	Global incidence is unknown; 300 to 500 cases in the U.S. every year

vade healthy cells. The antibody system works well against a number of disease-causing agents or pathogens—especially against the many bacteria that live outside a host's cells. In theory, antibodies could attack the microbes before they entered cells or when newly minted copies traveled from one cell to another. But the B lymphocyte system is not terribly effective at these tasks when it comes to chlamydiae, which live inside the cells, where circulating antibodies cannot reach them.

To prevent chlamydiae from lying dormant in cells and then proliferating anew, a vaccine would probably need to pump up the so-called cellular arm of the immune system in addition to evoking an antibody attack. This arm, critical to eradicating viruses (which also live inside cells), relies on killer and helper T cells as well as on scavenger cells known as macrophages to eliminate invaders. Unfortunately, even this trio of immune cells does an incomplete job of eliminating chlamydiae, too often allowing infected cells to survive and become bacteria-producing factories.

THE AUTHORS

DAVID M. OJCIUS, TONI DARVILLE and PATRIK M. BAVOIL each bring different expertise to chlamydia research. After studying the cellular and immunological aspects of infections for 12 years in France, Ojcius joined the faculty at the University of California, Merced, in 2004. Darville, who is a pediatric infectious disease specialist at the University of Arkansas for Medical Sciences, has explored the immunology of chlamydial infection since 1994 using mice and guinea pigs. As an associate professor at the University of Maryland, Baltimore, Bavoil works on the biochemistry and molecular biology of the disease.

Developing a vaccine able to evoke a better cellular response than the body could mount on its own is a tall order. Most existing vaccines elicit a targeted antibody response, but safely activating cellular immunity against many infectious diseases remains a challenging task. The job is particularly difficult in the case of chlamydiae because these bacteria have special ways of protecting themselves from attack by the cellular branch of the immune system.

Hidden Hijackers

LIKE CERTAIN OTHER bacterial pathogens, chlamydiae induce epithelial cells—in this case, those lining genital tracts, eyelids or lungs—to absorb them within a membrane-bound sac, or vacuole. Healthy cells typically attempt to kill internalized pathogens by having the entry vacuoles fuse with lysosomes, cellular structures containing enzymes that chop up proteins, lipids and DNA. All cells display the chopped-up pieces on proteins called major histocompatibility complex (MHC) molecules at the cell surface. Killer and helper T cells, which travel around the body continuously, will then glom on to MHC molecules that display bits of foreign proteins. If the T cells also receive other indications of trouble, they will deduce that the cells are infected and will orchestrate an attack on them.

But chlamydiae somehow compel their entry vacuoles to avoid lysosomes, enabling the bacteria to proliferate freely while separated physically from the rest of the infected cell. If the lysosomes cannot provide bits of the bacteria for display on the cell surface, patrolling T cells will not recognize that a cell harbors invaders. Understanding how the bacteria grow

CHLAMYDIA'S STEALTHY ATTACK

Sexually transmitted chlamydia leaves most of its victims unaware of their infections until the damage is irreversible. In the worst case, infection of a woman's fallopian tubes creates scar tissue that stops a fertilized egg from reaching the uterus (*main illustration*), leading to a life-threatening tubal (ectopic) pregnancy. New revelations about the bacteria's survival tactics (*insets*) may soon make it possible to interrupt chlamydia's silent attack [*see box on page 47*].

1 BACTERIA INVADE CELLS ...
Sporelike forms of chlamydiae known as elementary bodies invade cells lining the genital tract by forming a pit on the cell surface (*below*). Enclosed within a pinched-off piece of the cell's outer membrane (known as an entry vacuole), elementary bodies begin differentiating into noninfectious reticulate bodies. The bacteria thrive by extracting nutrients from the host cells' cytoplasm.

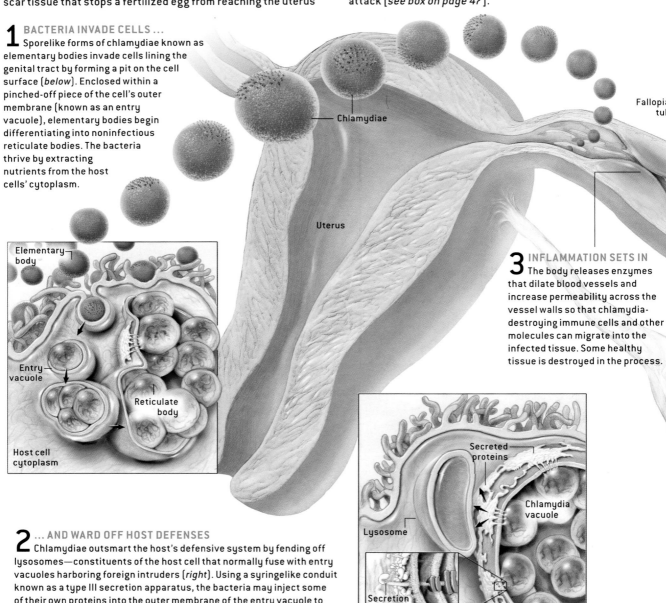

Chlamydiae

Fallopian tube

Uterus

Elementary body

Entry vacuole

Reticulate body

Host cell cytoplasm

3 INFLAMMATION SETS IN
The body releases enzymes that dilate blood vessels and increase permeability across the vessel walls so that chlamydia-destroying immune cells and other molecules can migrate into the infected tissue. Some healthy tissue is destroyed in the process.

Secreted proteins

Chlamydia vacuole

Lysosome

Secretion apparatus

2 ... AND WARD OFF HOST DEFENSES
Chlamydiae outsmart the host's defensive system by fending off lysosomes—constituents of the host cell that normally fuse with entry vacuoles harboring foreign intruders (*right*). Using a syringelike conduit known as a type III secretion apparatus, the bacteria may inject some of their own proteins into the outer membrane of the entry vacuole to physically block the lysosome's assault.

and avoid lysosomes might suggest new ways to forestall or halt the infection. Recent findings, including the newly sequenced chlamydial genomes, are aiding in that effort.

The sequence of genetic building blocks in an organism's DNA specifies the proteins that cells make; the proteins, in turn, carry out most cellular activities. Thus, the sequence of codes in a gene says a good deal about how an organism functions. Researchers, including Ru-ching Hsia and one of us (Bavoil) of the University of Maryland, discovered a particularly important element of chlamydiae by noting similarities between their genes and those of larger bacteria, such as *Salmonella typhimurium*, infamous for causing food poisoning. Scientists now generally agree that chlamydiae have everything they need to form a versatile, needlelike projection called a type III secretion apparatus. This apparatus, which spans the membrane of the entry vacuole, serves as a conduit between the bacteria and the cytoplasm of the host cell.

Such a connection implies that chlamydiae can inject pro-

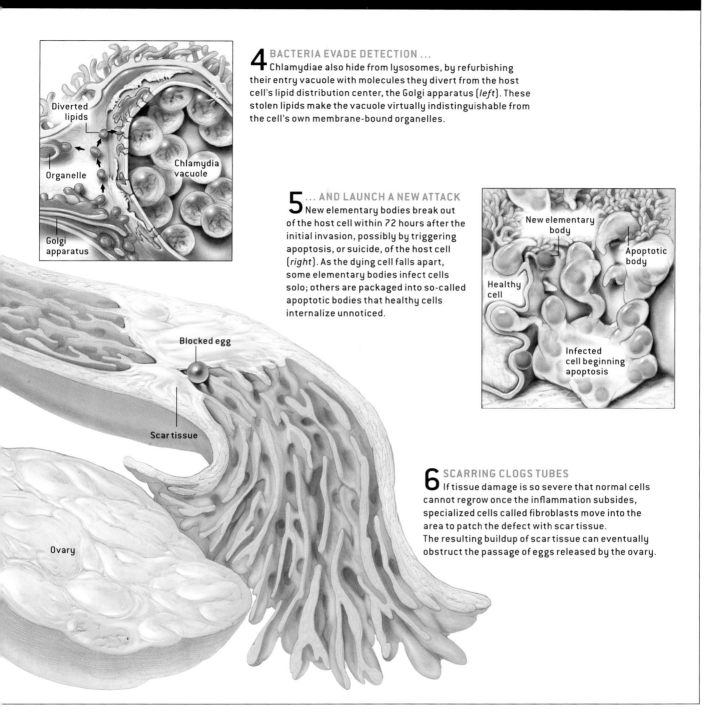

4 BACTERIA EVADE DETECTION ...
Chlamydiae also hide from lysosomes, by refurbishing their entry vacuole with molecules they divert from the host cell's lipid distribution center, the Golgi apparatus (*left*). These stolen lipids make the vacuole virtually indistinguishable from the cell's own membrane-bound organelles.

Diverted lipids

Organelle

Chlamydia vacuole

Golgi apparatus

5 ... AND LAUNCH A NEW ATTACK
New elementary bodies break out of the host cell within 72 hours after the initial invasion, possibly by triggering apoptosis, or suicide, of the host cell (*right*). As the dying cell falls apart, some elementary bodies infect cells solo; others are packaged into so-called apoptotic bodies that healthy cells internalize unnoticed.

New elementary body

Apoptotic body

Healthy cell

Infected cell beginning apoptosis

Blocked egg

Scar tissue

Ovary

6 SCARRING CLOGS TUBES
If tissue damage is so severe that normal cells cannot regrow once the inflammation subsides, specialized cells called fibroblasts move into the area to patch the defect with scar tissue.
The resulting buildup of scar tissue can eventually obstruct the passage of eggs released by the ovary.

teins into the cytoplasm of the host cell. The apparatus may thus help chlamydiae resist interaction with lysosomes, because it can secrete proteins that remodel the vacuole membrane in ways that bar lysosome function. In addition, investigators have watched the chlamydiae-bearing vacuole divert artificially fluorescing lipids from certain compartments of the host cell, including the Golgi apparatus, to the vacuole membrane. Normally, the membrane of an entry vacuole bears molecules made by the pathogen inside. In this case, a membrane

enclosing a bacterium would look foreign to the host cell, which would target the bacterium for immediate destruction by lysosomes. But the lipids that chlamydiae use to rebuild the membrane of their entry vacuole come from the host cell: the vacuoles are therefore indistinguishable from the host cell's organelles and invisible to lysosomes.

If scientists identify the proteins the bacteria secrete to camouflage vacuoles, they might be able to devise two kinds of infection-preventing treatments. One potential drug could interfere with the proteins' activity in a way that would force the entry vacuole to fuse with lysosomes, triggering an immune attack right after the chlamydiae invade the cell. Another drug might incapacitate the mechanisms the bacteria use to divert lipids from the host cell to the chlamydial vacuole, halting the trespassers' ability to hide. Hypothetically, such drugs could be incorporated into a topical microbicide that would thwart sexually transmitted chlamydiae.

Some of the proteins mentioned above—and any others that are unique to the bacteria and not made by human cells—might also be useful ingredients in vaccines. Newly sequenced genomes should be helpful in identifying good candidates.

Suicidal Tendencies

RECENT FINDINGS about the role of T cells may open other doors. Biologists have long known that killer T cells normally destroy infected cells by inducing a type of cell death known as apoptosis or "cell suicide," during which cells use their own enzymes to lyse their proteins and DNA. Also known is that immune cells—including T cells and macrophages—stimulate the production of cytokines that help to cripple bacteria and to trigger an inflammatory response that

stops their spread. One cytokine known to have this dual purpose is tumor necrosis factor-alpha (TNF-alpha). Laboratory investigations have shown, however, that some infected cells survive despite treatment with TNF-alpha and other apoptosis-inducing cytokines, leading to persistent infections. The problem is that the body does not give up easily. Cytokines continue to trigger chronic inflammation in an effort to contain the infection even if they cannot eliminate it outright.

But even persistently infected cells cannot live forever. Indeed, it appears that chlamydiae have developed their own way to elicit the death of a host cell, which they must do to ensure their own longevity. (The host cell must fall apart before the bacteria can infect other cells.) And as Jean-Luc Perfettini discovered while working as a graduate student with one of us (Ojcius) at the Pasteur Institute in Paris, chlamydiae can kill and exit the infected cells in a way that minimizes the host immune system's ability to sense any danger, thereby allowing the infection to spread essentially undetected in the body.

Addressing this final stage of the bacterial life cycle will require further investigation into the proteins involved in inducing apoptosis and in protecting persistently infected cells from suicidal signals. From what biologists know so far, the

Cardiovascular Connections

Colds, bronchitis and pneumonia may not be the sole concern for people who have inhaled the airborne species of *Chlamydophila*. Recent evidence hints that *C. pneumoniae* infections may also contribute to strokes and heart attacks. Such a link may have an upside, however—doctors might eventually be able to prescribe antibiotics to fight both the infection and the heart disease.

Atherosclerosis—a narrowing of the coronary arteries that leads to most strokes and heart attacks—causes approximately half of all adult deaths in the Western world. But traditional risk factors, such as elevated cholesterol and cigarette smoking, account for only about half of that total. Scientists searching for a reason behind the other 50 percent began to consider infections once it became clear that inflammation—a generalized immune response against any perceived invader—also underlies the growth and destructive ruptures of the fat-laden deposits that constrict coronary arteries [see "Atherosclerosis: The New View," by Peter Libby; SCIENTIFIC AMERICAN, May 2002].

C. pneumoniae became a prime suspect in the condition shortly after it was identified as a separate chlamydial species in 1983. It drew suspicion because of its ubiquity—more than 60 percent of adults worldwide carry antibodies against it (a sign of past or ongoing infection). Support for the hunch emerged in 1988, when physicians in Finland reported a positive correlation between the presence of these antibodies and the risk of developing coronary artery disease; other researchers identified the bacterium in clogged human

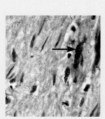

C. pneumoniae (*arrow*) turned up in the atherosclerotic coronary arteries of 54 percent of the 272 patients surveyed for a study published in 2000.

arteries five years later. Since then, organizations such as the National Institutes of Health and the American Heart Association have invested millions of dollars to study the relation between *C. pneumoniae* and atherosclerosis.

Animal studies conducted within the past five years or so have provided some of the most convincing evidence for a link. One demonstrated, for example, that chlamydial bacteria can move from the lungs of mice to other parts of the body within white blood cells, the agents responsible for inflammation. Other research has shown that *C. pneumoniae* infections accelerate atherosclerosis in both mice and rabbits and that antichlamydial antibiotics can prevent that acceleration.

Experimental results such as these, though tentative, were enough to justify a handful of small clinical trials in humans. Five of these trials showed that one to three months of antibiotic treatment had a statistically significant benefit against the progression of atherosclerosis. But results were mixed as to whether the antibiotics could actually prevent serious cardiac events. The promise for longer-term treatment was also dealt a blow by the negative outcomes of two large trials completed in 2004, each involving 4,000 volunteers who received antibiotics for one to two years.

Establishing whether a clear connection exists between *C. pneumoniae* infection and atherosclerosis in humans may prove difficult simply because so many other factors participate in heart disease. Exactly how troublesome such complications will be, however, remains to be seen. —*D.M.O., T.D. and P.M.B.*

CHLAMYDIA, INTERRUPTED

Sex education and improved hygiene cannot halt the spread of chlamydial infection on their own. That is why many scientists continue searching for an effective vaccine or other preventive treatments. Recent discoveries have suggested promising strategies, some listed below, for undermining the bacterium's survival tactics or limiting damage from an excessive immune response.

Kill the bacterium as it enters the body.

Develop a topical microbicide—a gel, cream or foam—that would be applied vaginally or rectally. Such products are in human trials to treat HIV, which infects the same tissues as chlamydia does.

Interfere with the bacterium's ability to invade the host cell.

Devise a vaccine that pumps up the host's antibody response. Following a vaccine or antibiotic treatment with anti-inflammatory drugs could decrease damage but has so far failed to do so in animal trials.

Inhibit the bacterium's growth within infected cells.

Interfere with the activity of the proteins chlamydia uses to divert lipids and other nutrients from the host cell. Such proteins have not yet been identified; once found, they could potentially be immobilized by a specially designed vaccine.

Promote intracellular destruction of the bacterium.

Disable the bacterium's type III secretion apparatus, which may release proteins that ward off lysosomes, constituents of the host cell that chop up foreign invaders.

In trials, bacteria with nearly identical apparatuses have been unable to cause symptoms of infection when scientists have disabled the genes that code for each apparatus; this finding suggests that drugs able to block the proteins encoded by those genes in chlamydia could be helpful.

Halt the bacterium's ability to spread.

Induce "suicide" of infected cells before the bacterium has a chance to convert into the form able to invade uninfected cells. Compounds that can induce premature cell death in tumors are under development; the same drugs could theoretically work against chlamydia.

---

latter avenue may prove more fruitful in developing a vaccine. By rendering persistently infected cells more sensitive to apoptosis, it might be possible to eliminate the bacteria that remain dormant in the system for long periods as well as decrease the lasting consequences of chronic infection.

Multiple Avenues of Attack

REGARDLESS OF THE DISCOVERIES that lie ahead, the ideal chlamydia vaccine will not be a simple one. It will have to activate both the antibody and cellular arms of the immune system more effectively than the body's natural response does yet somehow limit inflammation as well. For those concerned with preventing chlamydia-related STIs, an additional challenge is ensuring that memory lymphocytes remain in the genital tract poised to combat infection at all times. This tract does not contain the type of tissue that produces memory cells; such cells tend to vacate the area, leaving the person susceptible to infection after a brief period of immunity.

Recall that females bear the lasting effects of genital infection. One feasible goal of a vaccine might be to protect women from the disease rather than from infection per se. This aim might be achieved by vaccinating both men and women. In this scenario, the vaccine would have to generate only enough antibodies to reduce, rather than eliminate, the amount of bacteria men carry. Then, if a woman were exposed to a man's infection through intercourse, memory cells induced by her immunization would travel to the genital tract in numbers adequate for killing the relatively small number of organisms before they spread to her fallopian tubes.

Until researchers manage to develop such a vaccine, contraceptives that include antichlamydial drugs could pay off. These agents might take the form of compounds that either block the proteins chlamydiae use to bind to genital tract cells or target the proteins the microbes secrete to promote intracellular survival. For eye infections, the only vaccine likely to be useful is one that completely prevents infection.

While awaiting effective preventive strategies against chlamydia, it is worth remembering that current antibiotic treatment is highly successful when it is accessible. New details from genomic discoveries indicate that this efficacy will continue. Compared with free-living bacterial pathogens, which can share genes easily, the genomes of *Chlamydia* species have remained essentially the same for millions of years. This genetic stability implies that chlamydiae cannot easily acquire genes—including those for antibiotic resistance—from other bacteria.

It is also worth noting that antibiotics cannot undo the tissue damage caused by inflammation, and to be most useful, they must be given early. Therefore, more widespread screening of high-risk individuals is needed. Researchers have already proved the feasibility of employing noninvasive urine screening of sexually active young men and women, particularly in settings such as high schools, military intake centers and juvenile detention facilities. Public health officials need to pursue such strategies in parallel with the ongoing search for effective vaccines. SA

MORE TO EXPLORE

Chlamydia pneumoniae—An Infectious Risk Factor for Atherosclerosis? Lee Ann Campbell and Cho-cho Kuo in *Nature Reviews Microbiology*, Vol. 2, No. 1, pages 23–32; January 2004.

Chlamydia and Apoptosis: Life and Death Decisions of an Intracellular Pathogen. Gerald I. Byrne and David M. Ojcius in *Nature Reviews Microbiology*, Vol. 2, No. 10, pages 802–808; October 2004.

Host-Cell Survival and Death During Chlamydia Infection. Songmin Ying et al in *Current Immunology Reviews*, Vol. 3, No. 1, pages 33-40; February 2007.

Questions for Review

"Can Chlamydia Be Stopped"

By David M. Ojcius, Toni Darville, and Patrik M. Bavoil

TESTING YOUR COMPREHENSION

1. Which of the following is not correctly matched?
   a. *Chlamydophila pneumoniae* — exposure to aerosols
   b. *Chlamydophila psittaci* — contact with rodents
   c. *Chlamydia trachomatis* — transmitted by flies
   d. *Chlamydia trachomatis* — sexually transmitted

2. Chlamydial symptoms are due to
   a. bacterial toxins.
   b. host cell death.
   c. host immune response.
   d. bacterial waste products.

3. Chlamydia infections are difficult to treat because the bacteria
   a. are resistant to antibiotics.
   b. grow inside host cells.
   c. are usually fatal.
   d. are sexually transmitted.

4. *Chlamydia trachomatis* may be cultured in human cells. If infected cells were gram-stained what would you most likely observe at 1000×?
   a. Spores inside lysosomes
   b. A membrane-bounded inclusion body within the cytoplasm
   c. Gram-positive rods free within the cytoplasm
   d. Gram-negative cocci in the nucleus

5. Chlamydiae evade the host's immune system by all of the following except which one?
   a. Avoiding phagolysosome formation
   b. Building an entry vacuole of host cell lipids
   c. Causing secretion of cytokines
   d. Secreting lysosome-repelling proteins into the entry vacuole membrane

6. All of the following are characteristics of *Chlamydia trachomatis* except which one?
   a. They are obligate intracellular parasites.
   b. Tissue damage is caused primarily by immune reaction to repeated infection.
   c. The reticulate body is the infectious form.
   d. The host cell must fall apart before the bacteria can infect other cells.

7. The genus *Chlamydia* includes
   a. three species that are pathogenic to humans: *C. trachomatis*, *C. pneumoniae*, and *C. psittaci*.
   b. two species that are pathogenic to humans: *C. trachomatis* and *C. pneumoniae*.
   c. one species is pathogenic to humans: *C. trachomatis*.
   d. no species are human pathogens.

8. What property of chlamydiae enables them to change the chemical composition of the entry vacuole?
   a. Type III secretion system
   b. Flagella
   c. Endospore
   d. Elementary body

9. A type III secretion system enables chlamydia bacteria to export proteins. Which phrase best describes the process blocked by the type III secretion system?
   a. Cytokine synthesis by infected macrophages
   b. Fusion of entry vacuole with lysosomes
   c. Movement of neutrophils to the site of infection
   d. Processing and presentation of chlamydial antigens

10. Which of the following is not a realistic strategy for controlling chlamydiae?
    a. Believing a childhood infection will protect for life
    b. Disabling type III secretion
    c. Using a topical bactericidal gel
    d. Causing death of infected cells

MICROBIOLOGY IN SOCIETY

1. In 2006, nearly 1,000,000 cases of *Chlamydia trachomatis* were reported to the Centers of Disease Control and Prevention. Approximately 5%–14% of routinely screened females aged 16–24 years are infected with chlamydia. Approximately 30% of women aged 16–24 years are routinely screened, along with a pelvic examination and Pap smear, for chlamydia. What arguments would you make for routine screening? How would you use these statistics to reinforce your argument?

2. Fifteen percent of the world's blindness is due to trachoma. The economic burden of this due to years of productive life lost is estimated to be $29 billion annually. The cost of eradication is approximately $29 billion. Discuss reasons why trachoma is so common in developing countries but does not occur in developed countries. As a member of the World Health Organization, what measures, if any, would you implement to reduce the incidence of trachoma?

3. Should routine chlamydia screening be done in high school?

THINKING ABOUT MICROBIOLOGY

1. Vaccines are available against several bacterial infections. Why is it so difficult to make a vaccine against chlamydia?

2. What could be an advantage of *C. pneumonia* causing atherosclerosis?

3. Testing for chlamydia includes using a Gram stain to screen a urine sample for neutrophils. Positive samples are then tested with a DNA probe or polymerase chain reaction (PCR). Why not look for the bacteria in the Gram stain? What is the significance of a high number of neutrophils? How will a DNA probe or PCR diagnose the infection?

WRITING ABOUT MICROBIOLOGY

1. Discuss how chlamydial infections can lead to ectopic pregnancies and blindness.

2. Explain three ways that chlamydia evades the host's immune system.

3. Chlamydia bacteria are unusual because they have a life cycle consisting of two stages. Describe the life cycle of this bacterium.

Answers can be found on The Microbiology Place website. Go to www.microbiologyplace.com, click on the cover of your textbook, and type in your login name and password (using the access code found in the front pages of your textbook). Then, click on Current Issues Magazine Answers on the left navigation bar.

ATTACKING ANTHRAX

Recent discoveries are suggesting much-needed strategies for improving prevention and treatment. High on the list: ways to neutralize the anthrax bacterium's fiendish toxin

by John A. T. Young and R. John Collier

CULTURES OF CELLS survived exposure to the anthrax toxin after being treated with a potential antitoxin. Michael Moure of Harvard University holds a plate containing the culture

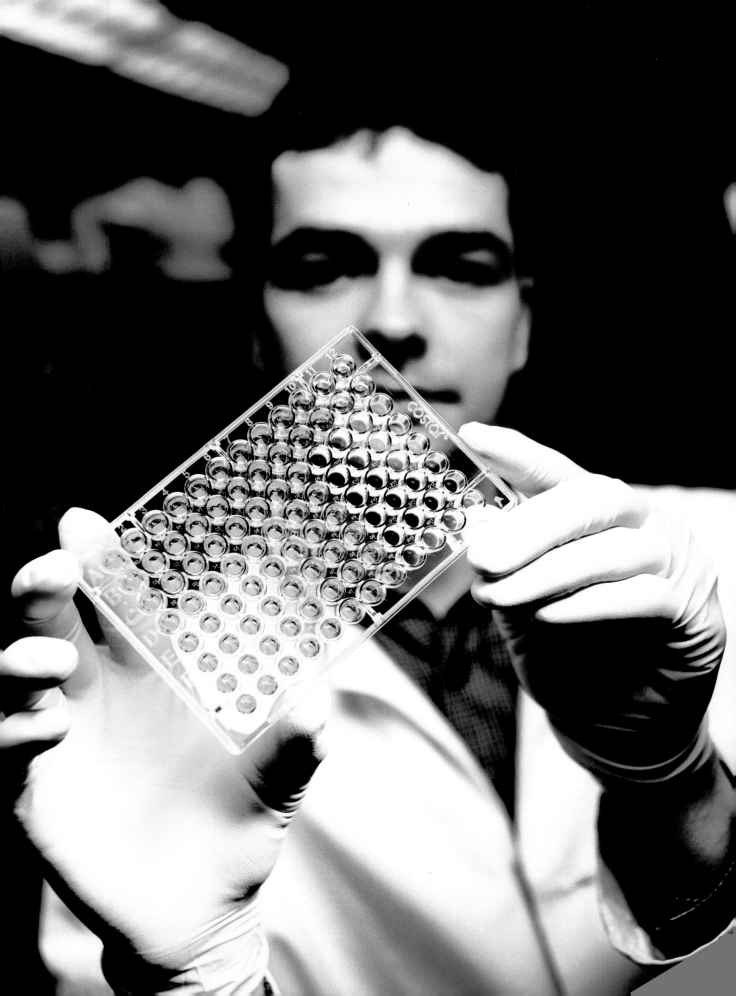

The need for new anthrax therapies became all too clear in 2001

when five people died of inhalation anthrax, victims of the first purposeful release of anthrax spores in the U.S. Within days of showing initially unalarming symptoms, the patients were gone, despite intensive treatment with antibiotics. Six others became seriously ill as well before pulling through.

Fortunately, our laboratories and others began studying the causative bacterium, *Bacillus anthracis,* and seeking antidotes long before fall 2001. Recent findings are now pointing the way to novel medicines and improved vaccines. Indeed, in the past few years, the two of us and our collaborators have reported on three promising drug prototypes.

An Elusive Killer

THE NEW IDEAS for fighting anthrax have emerged from ongoing research into how *B. anthracis* causes disease and death. Anthrax does not spread from individual to individual. A person (or animal) gets sick only after incredibly hardy spores enter the body through a cut in the skin, through contaminated food or through spore-laden air. Inside the body the spores molt into "vegetative," or actively dividing, cells.

Anthrax bacteria that colonize the skin or digestive tract initially do damage locally and may cause self-limited ailments: black sores and swelling in the first instance; possibly vomiting and abdominal pain and bleeding in the second. If bacterial growth persists unchecked in the skin or gastrointestinal tract, however, the microbes may eventually invade the bloodstream and thereby cause systemic disease.

Inhaled spores that reach deep into the lungs tend to waste little time where they land. They typically convert to the vegetative form and travel quickly to lymph nodes in the middle of the chest, where many of the cells find ready access to the blood. (Meanwhile bacteria that remain in the chest set the stage for a breath-robbing buildup of fluid around the lungs.)

Extensive replication in the blood is generally what kills patients who succumb to anthrax. *B. anthracis*'s ability to expand so successfully derives from its secretion of two substances, known as virulence factors, that can profoundly derail the immune defenses meant to keep bacterial growth in check. One of these factors encases the vegetative cells in a polymer capsule that inhibits ingestion by the immune system's macrophages and neutrophils—the scavenger cells that normally degrade disease-causing bacteria. The capsule's partner in crime is an extraordinary toxin that works its way into those scavenger cells, or phagocytes, and interferes with their usual bacteria-killing actions.

The anthrax toxin, which also enters other cells, is thought to contribute to mortal illness not only by dampening immune responses but also by playing a direct role. Evidence for this view includes the observation that the toxin alone, in the absence of bacteria, can kill animals. Conversely, inducing the immune system to neutralize the toxin prevents *B. anthracis* from causing disease.

A Terrible Toxin

HARRY SMITH and his co-workers at the Microbiological Research Establishment in Wiltshire, England, discovered the toxin in the 1950s. Aware of its central part in anthrax's lethality, many researchers have since focused on learning how the substance "intoxicates" cells—gets into them and disrupts their activities. Such details offer essential clues to blocking its effects. Stephen H. Leppla and Arthur M. Friedlander, while at the U.S. Army Medical Research Institute of Infectious Diseases, initiated that effort with their colleagues in the 1980s; the two

Overview/*Anthrax*

- A three-part toxin produced by the anthrax bacterium, *Bacillus anthracis,* contributes profoundly to the symptoms and lethality of anthrax.
- The toxin causes trouble only when it gets into the cytosol of cells, the material that bathes the cell's internal compartments.
- Drugs that prevented the toxin from reaching the cytosol would probably go a long way toward limiting illness and saving the lives of people infected by the anthrax bacterium.
- Analyses of how the toxin enters cells have recently led to the discovery of several potential antitoxins.

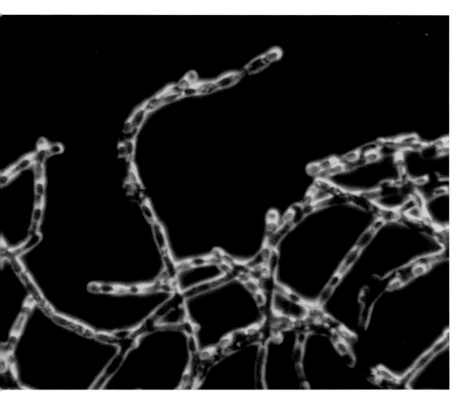

ACTIVELY DIVIDING CELLS of the anthrax bacterium arrange themselves into chains that resemble linked boxcars.

of us and others took up the task somewhat later.

The toxin turns out to consist of three proteins: protective antigen, edema factor and lethal factor. These proteins cooperate but are not always joined together physically. They are harmless individually until they attach to and enter cells, which they accomplish in a highly orchestrated fashion.

First, protective antigen binds to the surface of a cell, where an enzyme trims off its outermost tip. Next, seven of those trimmed molecules combine to form a ring-shaped structure, or heptamer, that captures the two factors and is transported to an internal membrane-bound compartment called an endosome. Mild acidity in this compartment causes the heptamer to change shape in a way that leads to the transport of edema factor and lethal factor across the endosomal membrane into the cytosol (the internal matrix of cells), where they do their mischief. In essence, the heptamer is like a syringe loaded with edema factor and lethal factor, and the slight acidity of the endosome causes the

syringe to pierce the membrane of the endosome and inject the toxic factors into the cytosol.

Edema factor and lethal factor catalyze different molecular reactions in cells. Edema factor upsets the controls on ion and water flow across cell membranes and thereby promotes the swelling of tissues. In phagocytes, it also saps energy that would otherwise be used to engulf bacteria.

The precise behavior of lethal factor, which could be more important in causing patient deaths, is less clear. Scientists do know that it is a protease (a protein-cutting enzyme) and that it cleaves enzymes in a family known as MAPKKs. Now they are trying to tease out the molecular events that follow such cleavage and to uncover the factor's specific contributions to disease and death.

Therapeutic Tactics

CERTAINLY DRUGS able to neutralize the anthrax toxin would help the immune system fight bacterial multiplication and would probably reduce a patient's risk of dying. At the moment, antibiotics given to victims of inhalation anthrax may control microbial expansion but leave the toxin free to wreak havoc.

In principle, toxin activity could be halted by interfering with any of the steps in the intoxication process. An attractive approach would stop the sequence almost before it starts, by preventing protective antigen from attaching to cells. Scientists realized almost 10 years ago that this protein initiated toxin entry by binding to some specific protein on the surface of cells; when cells were treated with enzymes that removed all their surface proteins, protective antigen found no footing. Until very recently, though, no one knew which of the countless proteins on cells served as the crucial receptor.

The two of us, with our colleagues Kenneth Bradley, Jeremy Mogridge and Michael Mourez, found the receptor in 2001. Detailed analysis of this molecule (now named ATR, for anthrax toxin receptor) then revealed that it spans the cell membrane and protrudes from it. The protruding part contains an area resembling a region that serves in other receptors as an attachment site for particular proteins. This discovery suggested that the area was the place where protective antigen latched onto ATR, and indeed it is.

We have not yet learned the normal function of the receptor, which surely did not evolve specifically to allow the anthrax toxin into cells. Nevertheless, knowledge of the molecule's makeup is enabling us to begin testing inhibitors of its activity. We have had success, for instance, with a compound called sATR, which is a soluble form of the receptor domain that binds to protective antigen. When sATR molecules are mixed into the medium surrounding cells, they serve as effective decoys, tricking protective antigen into binding to them instead of to its true receptor on cells.

THE AUTHORS

JOHN A. T. YOUNG and R. JOHN COLLIER have collaborated for several years on investigating the anthrax toxin. Young is Howard M. Temin Professor of Cancer Research in the McArdle Laboratory for Cancer Research at the University of Wisconsin–Madison. Collier, who has studied anthrax for more than 14 years, is Maude and Lillian Presley Professor of Microbiology and Molecular Genetics at Harvard Medical School.

Detecting Anthrax

Rapid sensing would save lives

By Rocco Casagrande

IF A TERRORIST GROUP spread anthrax spores into the open air, the release could affect large numbers of people but would probably go unnoticed until victims showed up at hospitals. Many would undoubtedly seek help too late to be saved by current therapies. Much illness could be prevented, however, if future defenses against anthrax attacks included sensors that raised an alarm soon after spores appeared in the environment. The needed instruments are not yet ready for deployment, but various designs that incorporate cutting-edge technology are being developed.

Environmental sensors must discriminate between disease-causing agents (pathogens) and the thousands of similar but harmless microorganisms that colonize air, water and soil. Most of the tools being investigated work by detecting unique molecules on the surface of the pathogens of interest or by picking out stretches of DNA found only in those organisms.

The Canary, which is being developed at the Massachusetts Institute of Technology Lincoln Laboratory, is an innovative example of the devices that detect pathogens based on unique surface molecules. The sensors of the Canary consist of living cells—B cells of the immune system—that have been genetically altered to emit light when their calcium levels change. Protruding from these cells are receptors that will bind only to a unique part of a surface molecule on a particular pathogen. When the cells in the sensor bind to their target, that binding triggers the release of calcium ions from stores within the cells, which in turn causes the cells to give off light. The Canary can discern more than one type of pathogen by running a sample through several cell-filled modules, each of which reacts to a selected microorganism.

The GeneXpert system, developed by Cepheid, in Sunnyvale, Calif., is an example of a gene-centered approach. It begins its work by extracting DNA from microorganisms in a sample. Then, if a pathogen of concern is present, small primers (strips of genetic material able to recognize specific short sequences of DNA) latch onto the ends of DNA fragments unique to the pathogen. Next, through a procedure called the polymerase chain reaction (PCR), the system makes many copies of the bound DNA, adding fluorescent labels to the new copies along the way. Within about 30 minutes GeneXpert can make enough DNA to reveal whether even a small amount of the worrisome organism inhabited the original sample.

This system contains multiple PCR reaction chambers with

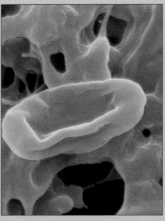

CARTRIDGE used in the experimental GeneXpert system is about as tall as an adult's thumb (*left*). Inside, sound waves bombard material to be tested, causing any cells to break open and release their DNA. If a pathogen of interest is present, its DNA will be amplified in the arrow-shaped reaction tube (*protrusion*), and the edges of the arrowhead will fluoresce. The micrograph (*right*) shows the remains of a cell that has disgorged its contents.

distinct primer sets to allow the detection of different pathogens simultaneously. Furthermore, the GeneXpert system could be used to determine whether the anthrax bacterium is present in a nasal swab taken from a patient in as little as half an hour, significantly faster than the time it takes for conventional microbiological techniques to yield results.

Instruments designed specifically to detect spores of the anthrax bacterium or of closely related microbes (such as the one that causes botulism) can exploit the fact that such spores are packed full of dipicolinic acid (DPA)—a compound, rarely found elsewhere in nature, that helps them to survive harsh environmental conditions. Molecules that fluoresce when bound to DPA have shown promise in chemically based anthrax detectors. "Electronic noses," such as the Cyranose detection system made by Cyrano Sciences in Pasadena, Calif., could possibly "smell" the presence of DPA in an air sample laced with anthrax spores.

The true danger of an anthrax release lies in its secrecy. If an attack is discovered soon after it occurs and if exposed individuals receive treatment promptly, victims have an excellent chance of surviving. By enhancing early detection, sensors based on the systems discussed above or on entirely different technologies could effectively remove a horrible weapon from a terrorist's arsenal.

ROCCO CASAGRANDE is a scientist at Surface Logix in Brighton, Mass., where he is developing methods and devices for detecting biological weapons.

We are now trying to produce sATR in the amounts needed for evaluating its ability to combat anthrax in rodents and nonhuman primates—experiments that must be done before any new drug can be considered for fighting anthrax in people. Other groups are examining whether carefully engineered antibodies (highly specific molecules of the immune system) might bind tightly to protective antigen in ways that will keep it from coupling with its receptor.

More Targets

SCIENTISTS ARE ALSO seeking ways to forestall later steps in the intoxication pathway. For example, a team from Harvard has constructed a drug able to clog the regions of the heptamer that grasp edema and lethal factors. The group—from the laboratories of one of us (Collier) and George M. Whitesides—reasoned that a plugged heptamer would be unable to draw the factors into cells.

We began by screening randomly constructed peptides (short chains of amino acids) to see if any of them bound to the

ALSO AVAILABLE!

Current Issues in Microbiology, Volume 1

2007 • Paper • ISBN-13: 978-0-8053-4623-7 • ISBN-10: 0-8053-4623-6

Give your students the best of both worlds—accessible, dynamic, and relevant articles from *Scientific American* magazine paired with the authority, reliability, and clarity of Benjamin Cummings microbiology textbooks.

Articles present key issues in microbiology, and end-of-article questions help students check their comprehension and make connections to science and society.

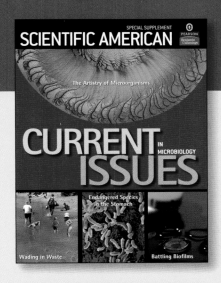

CONTENTS

1. Battling Biofilms
2. An Endangered Species in the Stomach
3. The Artistry of Microorganisms
4. Preparing for a Pandemic
5. The Science of Bad Breath
6. Intrigue at the Immune Synapse
7. Talking Bacteria
8. Wading in Waste

PEARSON
Benjamin Cummings

This engaging resource can be packaged with any Benjamin Cummings microbiology textbook. To request a complimentary copy, or to inquire about ordering options, please contact your Benjamin Cummings sales representative at **www.aw-bc.com/replocator**.

Microbiology, Ninth Edition
Tortora, Funke, Case

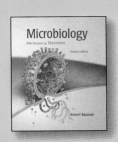

Microbiology with Diseases by Taxonomy, Second Edition
Bauman

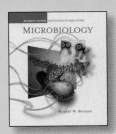

Microbiology with Diseases by Body System,
Bauman

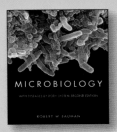

Microbiology with Diseases by Body System, Second Edition
Bauman
Coming in Spring 2008!

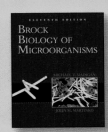

Brock Biology of Microorganisms, Eleventh Edition
Madigan, Martinko

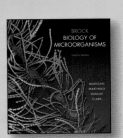

Brock Biology of Microorganisms, Twelfth Edition
Madigan, Martinko, Dunlap, Clark
Coming in Spring 2008!

ANTHRAX IN ACTION

Physicians classify anthrax according to the tissues that are initially infected. The disease turns deadly when the causative bacterium, *Bacillus anthracis,* reaches the bloodstream and proliferates there, producing large amounts of a dangerous toxin. Much research is now focused on neutralizing the toxin.

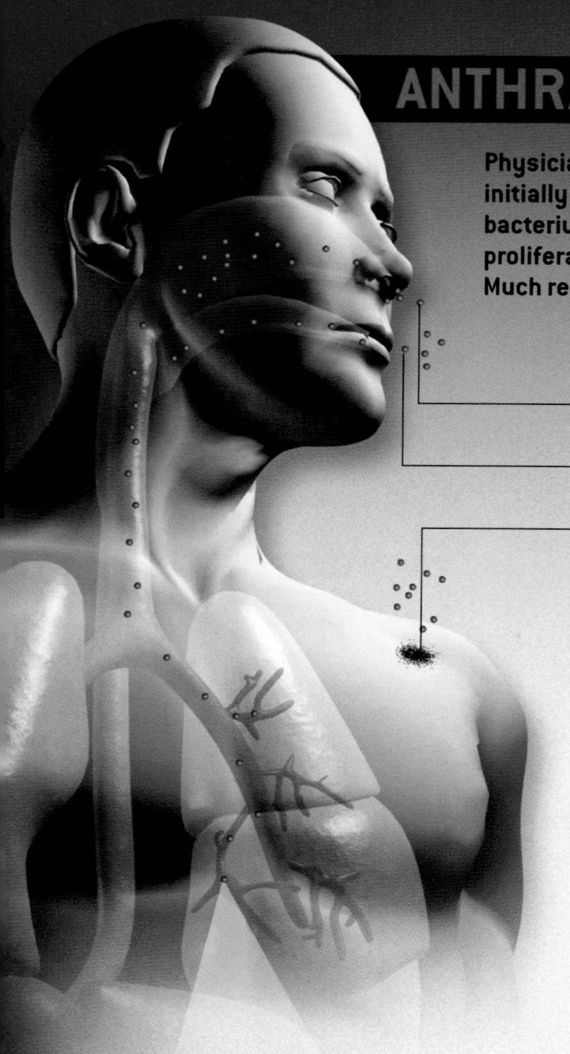

THREE TYPES

INHALATION ANTHRAX
Spores are breathed in

GASTROINTESTINAL ANTHRAX
Spores are ingested by eating contaminated meat

CUTANEOUS ANTHRAX
Spores penetrate the skin through a break

HOW INHALATION ANTHRAX ARISES

Inhalation anthrax is the most dangerous form, probably because bacteria that land in the lungs are more likely to reach the bloodstream and thus disseminate their toxin through the body.

1 Immune system cells called macrophages ingest *B. anthracis* spores and carry them to lymph nodes in the chest. En route, or in the macrophages, the spores transform into actively dividing cells

2 Proliferating *B. anthracis* cells erupt from macrophages and infiltrate the blood readily

3 In the blood, the active bacteria evade destruction by macrophages and other cells of the immune system by producing a capsule (*detail*) that blocks the immune cells from ingesting them and by producing a toxin that enters immune cells and impairs their functioning

4 Protected from immune destruction, the bacteria multiply freely and spread through the body

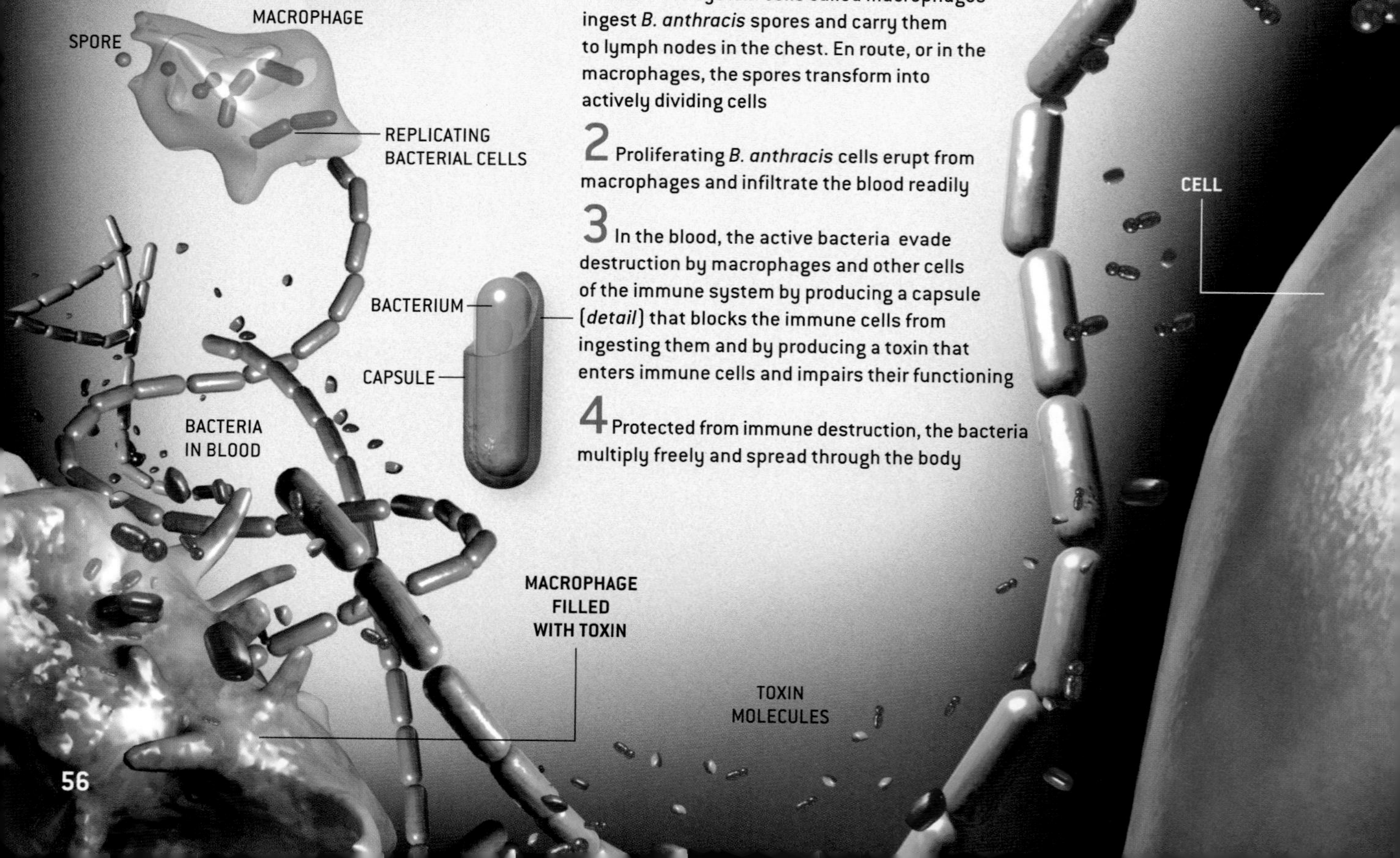

SPORE

MACROPHAGE

REPLICATING
BACTERIAL CELLS

BACTERIUM

CAPSULE

BACTERIA
IN BLOOD

MACROPHAGE
FILLED
WITH TOXIN

TOXIN
MOLECULES

CELL

HOW THE TOXIN INVADES CELLS ... AND HOW TO STOP IT

THE ANTHRAX TOXIN must enter cells to hurt the body. It consists of three collaborating proteins: protective antigen (PA), edema factor (EF) and lethal factor (LF). The last two disrupt cellular activities, but only after protective antigen delivers them to the cytosol—the matrix surrounding the cell's intracellular compartments. Molecular understanding of how the factors reach the cytosol has led to ideas for blocking that journey and thus for neutralizing the toxin and saving lives. The antitoxins depicted in the boxes have shown promise in laboratory studies.

HEPTAMER

4 Up to three copies of EF or LF or a combination of the two bind to the heptamer

3 Seven copies combine, forming a heptamer

2 PA gets cleaved

ANTHRAX TOXIN RECEPTOR (ATR)

PA binds to its ptor on a cell

PA

HEPTAMER COMPLEX

ENDOSOME

TREATMENT IDEA

EF   LF

INHIBITOR

Keep EF and LF from attaching to their binding sites on PA heptamers. Plug those sites with linked copies of a molecule that also has affinity for the sites.

5 The heptamer complexed with EF and LF is delivered to a membrane-bound compartment called an endosome

6 Mild acidity in the endosome causes the heptamer to inject EF and LF into the cytosol

TREATMENT IDEA

SOLUBLE RECEPTOR (sATR)

PA

Prevent PA from linking to its receptor on cells. Induce it to bind instead to decoys, such as soluble copies of the toxin receptor's PA binding site.

TREATMENT IDEA

ENDOSOME

HEPTAMER

DNI

CYTOSOL

Block transport of EF and LF from the endosome into the cytosol by causing newly forming heptamers to incorporate a version of PA known as a dominant negative inhibitor (DNI). DNI-containing heptamers cannot move EF and LF across the endosome's membrane.

CYTOSOL

8 LF is believed to be important in causing disease and death, but exactly how it does so is in question

7 EF causes tissues to swell and prevents immune system cells from ingesting and degrading bacteria

heptamer. One did, so we examined its ability to block toxin activity. It worked, but weakly. Assuming that fitting many plugs into the heptamer's binding domains for edema and lethal factor would be more effective, we took advantage of chemical procedures devised by Whitesides's group and linked an average of 22 copies of the peptide to a flexible polymer. That construction showed itself to be a strong inhibitor of toxin action—more than 7,000 times better than the free peptide—both in cell cultures and in rats.

And understanding of the receptor's three-dimensional structure would reveal the precise contact points between protective antigen and the receptor, enabling drugmakers to custom-design receptor blocking agents.

Scientists would also like to uncover the molecular interactions that enable protective antigen heptamers to move from the cell surface into endosomes inside the cell. Impeding that migration should be very useful. And what happens after lethal factor cleaves MAPKK en-

tralize the toxin of concern as soon as it appears in the body, thus preventing disease. Livestock in parts of the U.S. receive preparations consisting of *B. anthracis* cells that lack the protective capsule and thus replicate poorly. A similar vaccine for humans has been used in the former Soviet Union. But preparations that contain whole microbes often cause side effects, and they raise the specter that renegade cells might at times give rise to the very diseases they were meant to prevent.

The only anthrax vaccine approved

To be most effective, antitoxins will probably be USED WITH ANTIBIOTICS, much as cocktails of antiviral drugs are recommended for treating HIV infection.

Another exciting agent, and the one probably closest to human testing, would alter the heptamer itself. This compound was discovered after Bret R. Sellman in Collier's group noted that when certain mutant forms of protective antigen were mixed with normal forms, the heptamers formed on cells as usual but were unable to inject edema and lethal factors into the cytosol. Remarkably, some of these mutants were so disruptive that a single copy in a heptamer completely prevented injection.

In a study reported in April 2001, these mutants—known as dominant negative inhibitors, or DNIs—proved to be potent blockers of the anthrax toxin in cell cultures and in rats. Relatively small amounts of selected DNIs neutralized an amount of protective antigen and lethal factor that would otherwise kill a rat in 90 minutes. These findings suggest that each mutant copy of protective antigen is capable of inactivating six normal copies in the bloodstream and that it would probably reduce toxin activity in patients dramatically.

Of course, as more and more questions about the toxin are answered, scientists should discover further treatment ideas. Now that the receptor for protective antigen has been identified, researchers can use it as a target in screening tests aimed at finding drugs able to bar the receptor from binding to protective antigen.

zymes? How do those subsequent events affect cells? Although the latter question remains a vexing challenge, recent study of lethal factor has brightened the prospects for finding drugs able to inactivate it. In November 2001, Robert C. Liddington of the Burnham Institute in La Jolla, Calif., and his colleagues in several laboratories published the three-dimensional structure of the part of lethal factor that acts on MAPKK molecules. That site can now become a target for drug screening or design.

New leads for drugs should also emerge from the recent sequencing of the code letters composing the *B. anthracis* genome. By finding genes that resemble those of known functions in other organisms, biologists are likely to discover additional information about how the anthrax bacterium causes disease and how to stop it.

The continuing research should yield several antitoxins. To be most effective, such drugs will probably be used with antibiotics, much as cocktails of antiviral drugs are recommended for treating HIV infection.

Promising Preventives

AS PLANS TO IMPROVE therapies proceed, so does work on better vaccines. Vaccines against toxin-producing bacteria often prime the immune system to neu-

for human use in the U.S. takes a different form. It consists primarily of toxin molecules that have been chemically treated to prevent them from making people ill. It is produced by growing the weakened strain of *B. anthracis* in culture, filtering the bacterial cells from the culture medium, adsorbing the toxin proteins in the remaining filtrate onto an adjuvant (a substance that enhances immune responses) and treating the mixture with formaldehyde to inactivate the proteins. Injection of this preparation, known as AVA (for anthrax vaccine adsorbed), stimulates the immune system to produce antibodies that specifically bind to and inactivate the toxin's components. Most of the antibodies act on protective antigen, however, which explains the protein's name: it is the component that best elicits protective immunity.

AVA is given to soldiers and certain civilians but is problematic as a tool for shielding the general public against biological warfare. Supplies are limited. And even if AVA were available in abundance, it would be cumbersome to deliver on a large scale; the standard protocol calls for six shots delivered over 18 months followed by annual boosters. The vaccine has not been licensed for use in people already exposed to anthrax spores. But in late 2001 officials, worried that spores

Medical Lessons

Doctors now have a changed view of inhalation anthrax

By Ricki L. Rusting

NORMA WALLACE of Willingboro, N.J., is one of the six patients who survived inhalation anthrax in 2001

THE 2001 CASES of inhalation anthrax in the U.S. have upended some old assumptions about that disease. When contaminated letters started appearing in September 2001, public health authorities initially believed that only those who received the letters, and perhaps individuals nearby, were in danger. But spores clearly seeped out through the weave of the envelopes, contaminating postal facilities and jumping to other mail. Such "cross contamination" is a leading explanation for the deaths of two of the 11 people confirmed to have contracted inhalation anthrax in 2001. Also contrary to expectations, spores do not remain sedentary once they land. They can become airborne again as people walk around in a tainted room.

One surprise was positive. Before October 2001, common wisdom held that inhalation anthrax was almost always incurable after symptoms appeared. But doctors beat those odds in fall 2001, saving six of the victims. What made the difference? Researchers cannot draw firm conclusions from so few cases. But some intriguing patterns emerged when John A. Jernigan of the Centers for Disease Control and Prevention (CDC) and a team of others reviewed the medical records of the first 10 patients. Their findings appear in the November/December 2001 *Emerging Infectious Diseases* and online at www.cdc.gov/ncidod/eid/vol7no6/jernigan.htm

Relatively prompt diagnosis may have helped, the researchers report. Inhalation anthrax has two symptomatic phases—an early period marked by maladies common to a variety of ailments (such as fatigue, fever, aches and cough) and a later phase in which patients become critically ill with high fever, labored breathing and shock. Six of the 10 patients received antibiotics active against the anthrax bacterium, *Bacillus anthracis,* while they were still showing early symptoms of infection, and only they survived.

The types of antibiotics prescribed and the use of combinations of drugs might also have had a hand in the unexpectedly high survival rate. Nine of the people discussed in the review sought care before the CDC published what it called "interim" guidelines for treating inhalation anthrax on October 26, 2001, but most patients received therapy consistent with those guidelines: ciprofloxacin (the now famous Cipro) or doxycycline plus one or two other agents known to inhibit replication of *B. anthracis* (such as rifampin, vancomycin, penicillin, ampicillin, chloramphenicol, imipenem, clindamycin and clarithromycin). Aggressive "supportive" care—including draining dangerous fluid from around the lungs—probably helped as well, scientists say.

Even the survivors were very sick, however. Jernigan says they are still being observed to see whether long-term complications will develop, although as of mid-January, 2002, no obvious signs of such problems had emerged. Researchers suspect that anthrax antitoxins would ease the course of many people afflicted with anthrax and might also rescue patients who could not be saved with current therapies.

Ricki L. Rusting is a staff editor and writer.

might sometimes survive in the lungs for a long time, began offering an abbreviated, three-course dose on an experimental basis to postal workers and others who had already taken 60 days of precautionary antibiotics. People who accepted the offer were obliged to take antibiotics for an additional 40 days, after which the immunity stimulated by the vaccine would presumably be strong enough to provide adequate protection on its own.

In hopes of producing a more powerful, less cumbersome and faster-acting vaccine, many investigators are focusing on developing inoculants composed primarily of protective antigen produced by recombinant DNA technology. By coupling the recombinant protein with a potent new-generation adjuvant, scientists may be able to evoke good protective immunity relatively quickly with only one or two injections. The dominant negative inhibitors discussed earlier as possible treatments could be useful forms of protective antigen to choose. Those molecules retain their ability to elicit immune responses. Hence, they could do double duty: disarming the anthrax toxin in the short run while building up immunity that will persist later on.

We have no doubt that the expanding research on the biology of *B. anthracis* and on possible therapies and vaccines will one day provide a range of effective anthrax treatments. We fervently hope that these efforts will mean that nobody will have to die from anthrax acquired either naturally or as a result of biological terrorism. SA

MORE TO EXPLORE

Anthrax as a Biological Weapon: Medical and Public Health Management. Thomas V. Inglesby et al. in *Journal of the American Medical Association,* Vol. 281, No. 18, pages 1735–1745; May 12, 1999.

Dominant-Negative Mutants of a Toxin Subunit: An Approach to Therapy of Anthrax. Bret R. Sellman, Michael Mourez and R. John Collier in *Science,* Vol. 292, pages 695–697; April 27, 2001.

Designing a Polyvalent Inhibitor of Anthrax Toxin. Michael Mourez et al. in *Nature Biotechnology,* Vol. 19, pages 958–961; October 2001.

Identification of the Cellular Receptor for Anthrax Toxin. Kenneth A. Bradley, Jeremy Mogridge, Michael Mourez, R. John Collier and John A. T. Young in *Nature,* Vol. 414, pages 225–229; November 8, 2001.

The U.S. Centers for Disease Control and Prevention maintain a Web site devoted to anthrax at www.cdc.gov/ncidod/dbmd/diseaseinfo/anthrax_g.htm

Questions for Review

"Attacking Anthrax"

By John A. T. Young and R. John Collier

TESTING YOUR COMPREHENSION

1. The symptoms of anthrax are due to
   a. a bacterial endotoxin.
   b. a bacterial exotoxin.
   c. a hyperimmune response.
   d. bacterial growth in the body.

2. Which one of the following is not a virulence factor for *Bacillus anthracis*?
   a. Capsule
   b. Cell wall
   c. Edema factor
   d. Protective antigen

3. EF works by causing ions to move
   a. into a bacterial cell.
   b. into a host cell.
   c. out of a bacterial cell.
   d. out of a host cell.

4. The following steps occur during anthrax intoxication. Which is the second step?
   a. Edema factor (EF) binds to protective antigen (PA)
   b. Lethal factor (LF) binds to PA
   c. PA forms heptamer
   d PA is cleaved

5. Anthrax can be transmitted by all of the following mechanisms except which one?
   a. Ingesting endospores
   b. Inhaling endospores
   c. Inhaling any one of the anthrax toxins
   d. Endospores entering through cut skin

6. Which of the following is responsible for moving anthrax toxin into the cytosol?
   a. Dominant negative inhibitor (DNI)
   b. EF
   c. Endosome
   d. PA

7. The method of action of DNI molecules is to
   a. kill bacteria.
   b. link with a PA heptamer.
   c. neutralize toxin.
   d. prevent endospore germination.

8. The anthrax toxin binds human cells at
   a. ATR.
   b. DNI.
   c. LF.
   d. PA.

9. AVA contains
   a. anthrax toxins and an adjuvant.
   b. *B. anthracis* cells.
   c. *B. anthracis* endospores.
   d. a medium in which *B. anthracis* has grown.

10. All of the following are rapid detection systems for *B. anthracis* except which one?
    a. B-cells that emit light in the presence of *B. anthracis*
    b. Copies of *B. anthracis* DNA that fluoresce
    c. Fluorescence of dipicolinic acid
    d. Modification of ATR to sATR

MICROBIOLOGY IN SOCIETY

1. Explain why it is not uncommon to vaccinate cattle against anthrax but rare to vaccinate humans.

2. Discuss the advantages and disadvantages of the following types of vaccines: dead *B. anthracis* cells, *B. anthracis* cells that lack capsules, and a DNA vaccine.

3. Cases of inhalation anthrax have occurred in the United States in people who have purchased imported goat-skin drums and yak yarn macramé. The source countries in Africa, the Caribbean, and the Middle East are some of the poorest countries on Earth and the people are trying to make a living. Describe your policy to prevent these imported occurrences.

THINKING ABOUT MICROBIOLOGY

1. Human anthrax is rare in the United States. A 2006 Pennsylvania man was the first case of naturally acquired inhalation anthrax since 1976. Between 1955 and 1999, 153 cases occurred during industrial processing of animals hides. What is the natural habitat of *B. anthracis*?

2. What are the virulence factors of *B. anthracis* and how do they lead to disease? How does this bacterium survive if it kills its host?

3. The symptoms of anthrax are due to intoxication and antibiotics are used to treat infections. Of what value are antibiotics for treating anthrax?

WRITING ABOUT MICROBIOLOGY

1. Describe how the three-part toxin of *B. anthracis* works.

2. Describe the three methods of contracting anthrax.

3. What is the value of administering an antitoxin and an antibiotic?

Answers can be found on The Microbiology Place website. Go to www.microbiologyplace.com, click on the cover of your textbook, and type in your login name and password (using the access code found in the front pages of your textbook). Then, click on Current Issues Magazine Answers on the left navigation bar.

The Challenge of Antibiotic Resistance

Certain bacterial infections now defy all antibiotics. The resistance problem may be reversible, but only if society begins to consider how the drugs affect "good" bacteria as well as "bad"

by Stuart B. Levy

In 2002 an event doctors had been fearing finally occurred. A patient in the United States with an often deadly bacterium, *Staphylococcus aureus*, did not respond to a once reliable antidote — the antibiotic vancomycin. Fortunately, in this patient, the staph microbe remained susceptible to other drugs and was eradicated. But the appearance of *S. aureus* not readily cleared by vancomycin foreshadows trouble.

Worldwide, many strains of *S. aureus* are already resistant to all antibiotics except vancomycin. Emergence of forms lacking sensitivity to vancomycin signifies that variants untreatable by every known antibiotic are on their way. *S.* *aureus*, a major cause of hospital-acquired infections, has thus moved one step closer to becoming an unstoppable killer.

The looming threat of incurable *S. aureus* is just the latest twist in an international public health nightmare: increasing bacterial resistance to many antibiotics that once cured bacterial diseases readily. Ever since antibiotics became widely available in the 1940s, they have been hailed as miracle drugs— magic bullets able to eliminate bacteria without doing much harm to the cells of treated individuals. Yet with each passing decade, bacteria that defy not only single but multiple antibiotics—and therefore are extremely difficult to con-

trol—have become increasingly common.

What is more, strains of at least three bacterial species capable of causing life-threatening illnesses (*Enterococcus faecalis, Mycobacterium tuberculosis* and *Pseudomonas aeruginosa*) already evade every antibiotic in the clinician's armamentarium, a stockpile of more than 100 drugs. In part because of the rise in resistance to antibiotics, the death rates for some communicable diseases (such as tuberculosis) have started to rise again, after having declined in the industrial nations.

How did we end up in this worrisome, and worsening, situation? Several interacting processes are at fault. Analyses of them point to a number of actions that

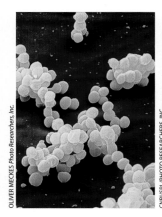

Staphylococcus aureus

Causes blood poisoning, wound infections and pneumonia; in some hospitals, more than 60 percent of strains are resistant to methicillin; some are poised for resistance to all antibiotics (H/C; 1950s)

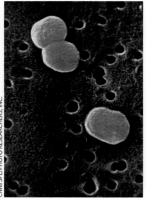

Acinetobacter

Causes blood poisoning in patients with compromised immunity (H, 1990s)

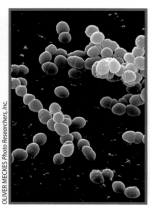

Enterococcus faecalis

Causes blood poisoning and urinary tract and wound infections in patients with compromised immunity; some multidrug-resistant strains are untreatable (H, 1980s)

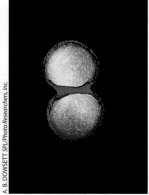

Neisseria gonorrhoeae

Causes gonorrhea; multidrug resistance now limits therapy chiefly to cephalosporins (C; 1970s)

Haemophilus influenzae

Causes pneumonia, e infections and mening especially in children. largely preventable vaccines (C; 1970s)

could help reverse the trend, if individuals, businesses and governments around the world can find the will to implement them.

One component of the solution is recognizing that bacteria are a natural, and needed, part of life. Bacteria, which are microscopic, single-cell entities, abound on inanimate surfaces and on parts of the body that make contact with the outer world, including the skin, the mucous membranes and the lining of the intestinal tract. Most live blamelessly. In fact, they often protect us from disease, because they compete with, and thus limit the proliferation of, pathogenic bacteria—the minority of species that can multiply aggressively (into the millions) and damage tissues or otherwise cause illness. The benign competitors can be important allies in the fight against antibiotic-resistant pathogens.

People should also realize that although antibiotics are needed to control bacterial infections, they can have broad, undesirable effects on microbial ecology. That is, they can produce long-lasting change in the kinds and proportions of bacteria—and the mix of antibiotic-resistant and antibiotic-susceptible types—not only in the treated individual but also in the environment and society at large. The compounds should thus be used only when they are truly needed, and they should not be administered for viral infections, over which they have no power.

A Bad Combination

Although many factors can influence whether bacteria in a person or in a community will become insensitive to an antibiotic, the two main forces are the prevalence of resistance genes (which give rise to proteins that shield bacteria from an antibiotic's effects) and the extent of antibiotic use. If the collective bacterial flora in a community have no genes conferring resistance to a given antibiotic, the antibiotic will successfully eliminate infection caused by any of the bacterial species in the collection. On the other hand, if the flora possess resistance genes and the community uses the drug persistently, bacteria able to defy eradication by the compound will emerge and multiply.

Antibiotic-resistant pathogens are not more virulent than susceptible ones: the same numbers of resistant and susceptible bacterial cells are required to produce disease. But the resistant forms are harder to destroy. Those that are slightly insensitive to an antibiotic can often be eliminated by using more of the drug; those that are highly resistant require other therapies.

To understand how resistance genes enable bacteria to survive an attack by an antibiotic, it helps to know exactly what antibiotics are and how they harm bacteria. Strictly speaking, the compounds are defined as natural substances (made by living organisms) that inhibit the growth, or proliferation, of bacteria or kill them directly. In practice, though, most commercial antibiotics have been chemically altered in the laboratory to improve their potency or to increase the range of species they affect. Here I will also use the term to encompass completely synthetic medicines, such as quinolones and sulfonamides, which technically fit under the broader rubric of antimicrobials.

Whatever their monikers, antibiotics, by inhibiting bacterial growth, give a host's immune defenses a chance to outflank the bugs that remain. The drugs typically retard bacterial proliferation by entering the microbes and interfering with the production of components needed to form new bacterial cells. For instance, the antibiotic tetracycline binds to ribosomes (internal structures that make new proteins) and, in so doing, impairs protein manufacture; penicillin and vancomycin impede proper synthesis of the bacterial cell wall.

Certain resistance genes ward off destruction by giving rise to enzymes that

ROGUE'S GALLERY OF BACTERIA features some types having variants resistant to multiple antibiotics; multidrug-resistant bacteria are difficult and expensive to treat. Certain strains of the species described in red no longer respond to any antibiotics and produce incurable infections. Some of the bacteria cause infections mainly in hospitals (H) or mainly in the community (C); others, in both settings. The decade listed with each entry indicates the period when resistance first became a significant problem for patient care. The bacteria, which are microscopic, are highly magnified in these false-color images.

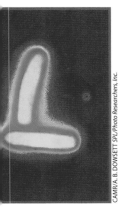

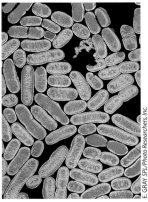

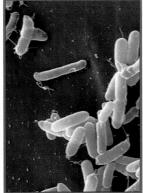

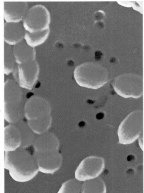

Mycobacterium tuberculosis

Causes tuberculosis; the multidrug-resistant strains are untreatable (H/C; 1970s)

Escherichia coli

Causes urinary tract infections, blood poisoning, diarrhea and kidney failure; some strains that cause urinary tract infections are multidrug-resistant (H/C; 1960s)

Pseudomonas aeruginosa

Causes blood poisoning and pneumonia, especially in people with cystic fibrosis or compromised immunity; some multidrug-resistant strains are untreatable (H/C; 1960s)

Shigella dysenteriae

Causes dysentery (bloody diarrhea); resistant strains have led to epidemics, and some can be treated only by expensive fluoroquinolones, which are often unavailable in developing nations (C; 1960s)

Streptococcus pneumoniae

Causes blood poisoning, middle ear infections, pneumonia and meningitis (C; 1970s)

ANTIBIOTIC-RESISTANT BACTERIA owe their drug insensitivity to resistance genes. For example, such genes might code for "efflux" pumps that eject antibiotics from cells (*a*). Or the genes might give rise to enzymes that degrade the antibiotics (*b*) or that chemically alter—and inactivate—the drugs (*c*). Resistance genes can reside on the bacterial chromosome or, more typically, on small rings of DNA called plasmids. Some of the genes are inherited, some emerge through random mutations in bacterial DNA, and some are imported from other bacteria.

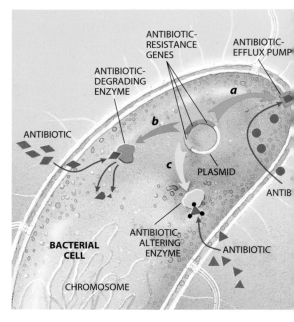

degrade antibiotics or that chemically modify, and so inactivate, the drugs. Alternatively, some resistance genes cause bacteria to alter or replace molecules that are normally bound by an antibiotic—changes that essentially eliminate the drug's targets in bacterial cells. Bacteria might also eliminate entry ports for the drugs or, more effectively, may manufacture pumps that export antibiotics before the medicines have a chance to find their intracellular targets.

My Resistance Is Your Resistance

Bacteria can acquire resistance genes through a few routes. Many inherit the genes from their forerunners. Other times, genetic mutations, which occur readily in bacteria, will spontaneously produce a new resistance trait or will strengthen an existing one. And frequently, bacteria will gain a defense against an antibiotic by taking up resistance genes from other bacterial cells in the vicinity. Indeed, the exchange of genes is so pervasive that the entire bacterial world can be thought of as one huge multicellular organism in which the cells interchange their genes with ease.

Bacteria have evolved several ways to share their resistance traits with one another [see "Bacterial Gene Swapping in Nature," by Robert V. Miller; SCIENTIFIC AMERICAN, January]. Resistance genes commonly are carried on plasmids, tiny loops of DNA that can help bacteria survive various hazards in the environment. But the genes may also occur on the bacterial chromosome, the larger DNA molecule that stores the genes needed for the reproduction and routine maintenance of a bacterial cell.

The Antibacterial Fad: A New Threat

Antibiotics are not the only antimicrobial substances being overexploited today. Use of antibacterial agents—compounds that kill or inhibit bacteria but are too toxic to be taken internally—has been skyrocketing as well. These compounds, also known as disinfectants and antiseptics, are applied to inanimate objects or to the skin.

Historically, most antibacterials were used in hospitals, where they were incorporated into soaps and surgical clothes to limit the spread of infections. More recently, however, those substances (including triclocarbon, triclosan and such quaternary ammonium compounds as benzalkonium chloride) have been mixed into soaps, lotions and dishwashing detergents meant for general consumers. They have also been impregnated into such items as toys, high chairs, mattress pads and cutting boards.

There is no evidence that the addition of antibacterials to such household products wards off infection. What is clear, however, is that the proliferation of products containing them raises public health concerns.

Like antibiotics, antibacterials can alter the mix of bacteria: they simultaneously kill susceptible bacteria and promote the growth of resistant strains. These resistant microbes may include bacteria that were present from the start. But they can also include ones that were unable to gain a foothold previously and are now able to thrive thanks to the destruction of competing microbes. I worry particularly about that second group—the interlopers—because once they have a chance to proliferate, some may become new agents of disease.

The potential overuse of antibacterials in the home is troubling on other grounds as well. Bacterial genes that confer resistance to antibacterials are sometimes carried on plasmids (circles of DNA) that also bear antibiotic-resistance genes. Hence, by promoting the growth of bacteria bearing such plasmids, antibacterials may actually foster double resistance—to antibiotics as well as antibacterials.

Routine housecleaning is surely necessary. But standard soaps and detergents (without added antibacterials) decrease the numbers of potentially troublesome bacteria perfectly well. Similarly, quickly evaporating chemicals—such as the old standbys of chlorine bleach, alcohol, ammonia and hydrogen peroxide—can be applied beneficially. They remove potentially disease-causing bacteria from, say, thermometers or utensils used to prepare raw meat for cooking, but they do not leave long-lasting residues that will continue to kill benign bacteria and increase the growth of resistant strains long after target pathogens have been removed.

If we go overboard and try to establish a sterile environment, we will find ourselves cohabiting with bacteria that are highly resistant to antibacterials and, possibly, to antibiotics. Then, when we really need to disinfect our homes and hands—as when a family member comes home from a hospital and is still vulnerable to infection—we will encounter mainly resistant bacteria. It is not inconceivable that with our excessive use of antibacterials and antibiotics, we will make our homes, like our hospitals, havens of ineradicable disease-producing bacteria.   —S.B.L.

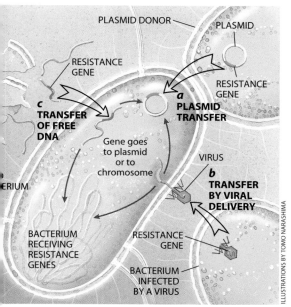

PLASMID DONOR

RESISTANCE GENE

PLASMID

RESISTANCE GENE

c TRANSFER OF FREE DNA

a PLASMID TRANSFER

Gene goes to plasmid or to chromosome

...ERIUM

VIRUS

b TRANSFER BY VIRAL DELIVERY

BACTERIUM RECEIVING RESISTANCE GENES

RESISTANCE GENE

BACTERIUM INFECTED BY A VIRUS

ILLUSTRATIONS BY TOMO NARASHIMA

BACTERIA PICK UP RESISTANCE GENES from other bacterial cells in three main ways. Often they receive whole plasmids bearing one or more such genes from a donor cell (*a*). Other times, a virus will pick up a resistance gene from one bacterium and inject it into a different bacterial cell (*b*). Alternatively, bacteria sometimes scavenge gene-bearing snippets of DNA from dead cells in their vicinity (*c*). Genes obtained through viruses or from dead cells persist in their new owner if they become incorporated stably into the recipient's chromosome or into a plasmid.

Often one bacterium will pass resistance traits to others by giving them a useful plasmid. Resistance genes can also be transferred by viruses that occasionally extract a gene from one bacterial cell and inject it into a different one. In addition, after a bacterium dies and releases its contents into the environment, another will occasionally take up a liberated gene for itself.

In the last two situations, the gene will survive and provide protection from an antibiotic only if integrated stably into a plasmid or chromosome. Such integration occurs frequently, though, because resistance genes are often embedded in small units of DNA, called transposons, that readily hop into other DNA molecules. In a regrettable twist of fate for human beings, many bacteria play host to specialized transposons, termed integrons, that are like flypaper in their propensity for capturing new genes. These integrons can consist of several different resistance genes, which are passed to other bacteria as whole regiments of antibiotic-defying guerrillas.

Many bacteria possessed resistance genes even before commercial antibiotics came into use. Scientists do not know exactly why these genes evolved and were maintained. A logical argument holds that natural antibiotics were initially elaborated as the result of chance genetic mutations. Then the compounds, which turned out to eliminate competitors, enabled the manufacturers to survive and proliferate—if they were also lucky enough to possess genes that protected them from their own chemical weapons. Later, these protective genes found their way into other species, some of which were pathogenic.

Regardless of how bacteria acquire resistance genes today, commercial antibiotics can select for—promote the survival and propagation of—antibiotic-resistant strains. In other words, by encouraging the growth of resistant pathogens, an antibiotic can actually contribute to its own undoing.

How Antibiotics Promote Resistance

The selection process is fairly straightforward. When an antibiotic attacks a group of bacteria, cells that are highly susceptible to the medicine will die. But cells that have some resistance from the start, or that acquire it later (through mutation or gene exchange), may survive, especially if too little drug is given to overwhelm the cells that are present. Those cells, facing reduced competition from susceptible bacteria, will then go on to proliferate. When confronted with an antibiotic, the most resistant cells in a group will inevitably outcompete all others.

Promoting resistance in known pathogens is not the only self-defeating activity of antibiotics. When the medicines attack disease-causing bacteria, they also affect benign bacteria—innocent bystanders—in their path. They eliminate drug-susceptible bystanders that could otherwise limit the expansion of pathogens, and they simultaneously encourage the growth of resistant bystanders. Propagation of these resistant, nonpathogenic bacteria increases the reservoir of resistance traits in the bacterial population as a whole and raises the odds that such traits will spread to pathogens. In addition, sometimes the growing populations of bystanders themselves become agents of disease.

Widespread use of cephalosporin antibiotics, for example, has promoted the proliferation of the once benign intestinal bacterium *E. faecalis*, which is naturally resistant to those drugs. In most people, the immune system is able to check the growth of even multidrug-resistant *E. faecalis,* so that it does not produce illness. But in hospitalized patients with compromised immunity, the enterococcus can spread to the heart valves and other organs and establish deadly systemic disease.

Moreover, administration of vancomycin over the years has turned *E. faecalis* into a dangerous reservoir of vancomycin-resistance traits. Recall that some strains of the pathogen *S. aureus*

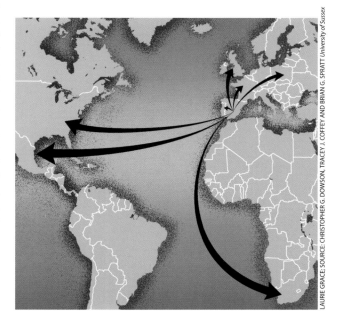

LAURIE GRACE; SOURCE: CHRISTOPHER G. DOWSON, TRACEY J. COFFEY AND BRIAN G. SPRATT *University of Sussex*

SPREAD OF RESISTANT BACTERIA, which occurs readily, can extend quite far. In one example, investigators traced a strain of multidrug-resistant *Streptococcus pneumoniae* from Spain to Portugal, France, Poland, the U.K., South Africa, the U.S. and Mexico.

are multidrug-resistant and are responsive only to vancomycin. Because vancomycin-resistant *E. faecalis* has become quite common, public health experts fear that it will soon deliver strong vancomycin resistance to those *S. aureus* strains, making them incurable.

The bystander effect has also enabled multidrug-resistant strains of *Acinetobacter* and *Xanthomonas* to emerge and become agents of potentially fatal bloodborne infections in hospitalized patients. These formerly innocuous microbes were virtually unheard of just ten years ago.

As I noted earlier, antibiotics affect the mix of resistant and nonresistant bacteria both in the individual being treated and in the environment. When resistant bacteria arise in treated individuals, these microbes, like other bacteria, spread readily to the surrounds and to new hosts. Investigators have shown that when one member of a household chronically takes an antibiotic to treat acne, the concentration of antibiotic-resistant bacteria on the skin of family members rises. Similarly, heavy use of antibiotics in such settings as hospitals, day care centers and farms (where the drugs are often given to livestock for nonmedicinal purposes) increases the levels of resistant bacteria in people and other organisms who are not being treated—including in individuals who live near those epicenters of high consumption or who pass through the centers.

Given that antibiotics and other antimicrobials, such as fungicides, affect the kinds of bacteria in the environment and people around the individual being treated, I often refer to these substances as societal drugs—the only class of therapeutics that can be so designated. Anticancer drugs, in contrast, affect only the person taking the medicines.

On a larger scale, antibiotic resistance that emerges in one place can often spread far and wide. The ever increasing volume of international travel has hastened transfer to the U.S. of multidrug-resistant tuberculosis from other countries. And investigators have documented the migration of a strain of multidrug-resistant *Streptococcus pneumoniae* from Spain to the U.K., the U.S., South Africa and elsewhere. This bacterium, also known as the pneumococcus, is a cause of pneumonia and meningitis, among other diseases.

Antibiotic Use Is Out of Control

For those who understand that antibiotic delivery selects for resistance, it is not surprising that the international community currently faces a major public health crisis. Antibiotic use (and misuse) has soared since the first commercial versions were introduced and now includes many nonmedicinal applications. In 1954 two million pounds were produced in the U.S.; today the figure exceeds 50 million pounds.

Human treatment accounts for roughly half the antibiotics consumed every year in the U.S. Perhaps only half that use is appropriate, meant to cure bacterial infections and administered correctly—in ways that do not strongly encourage resistance.

Notably, many physicians acquiesce to misguided patients who demand antibiotics to treat colds and other viral infections that cannot be cured by the drugs. Researchers at the Centers for Disease Control and Prevention have estimated that some 50 million of the 150 million outpatient prescriptions for antibiotics every year are unneeded. At a seminar I conducted, more than 80 percent of the physicians present admitted to having written antibiotic prescriptions on demand against their better judgment.

In the industrial world, most antibiotics are available only by prescription, but this restriction does not ensure proper use. People often fail to finish the full course of treatment. Patients then stockpile the leftover doses and medicate themselves, or their family and friends, in less than therapeutic amounts. In both circumstances, the improper dosing will fail to eliminate the disease agent completely and will, furthermore,

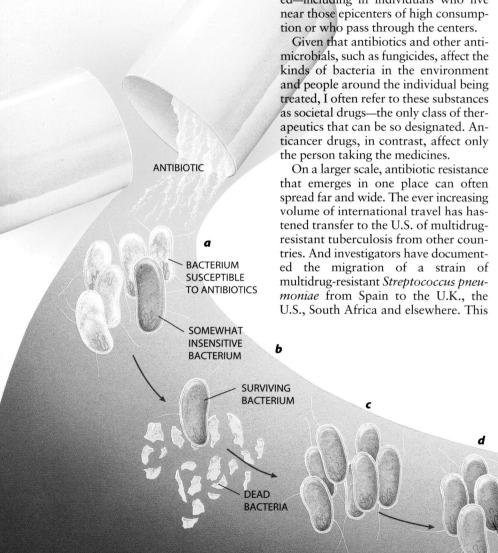

ANTIBIOTIC

a

BACTERIUM SUSCEPTIBLE TO ANTIBIOTICS

SOMEWHAT INSENSITIVE BACTERIUM

b

SURVIVING BACTERIUM

c

d

e

DEAD BACTERIA

BACTERIUM WITH INCREASED RESISTANCE

TOMO NARASHIMA

encourage growth of the most resistant strains, which may later produce hard-to-treat disorders.

In the developing world, antibiotic use is even less controlled. Many of the same drugs marketed in the industrial nations are available over the counter. Unfortunately, when resistance becomes a clinical problem, those countries, which often do not have access to expensive drugs, may have no substitutes available.

The same drugs prescribed for human therapy are widely exploited in animal husbandry and agriculture. More than 40 percent of the antibiotics manufactured in the U.S. are given to animals. Some of that amount goes to treating or preventing infection, but the lion's share is mixed into feed to promote growth. In this last application, amounts too small to combat infection are delivered for weeks or months at a time. No one is entirely sure how the drugs support growth. Clearly, though, this long-term exposure to low doses is the perfect formula for selecting increasing numbers of resistant bacteria in the treated animals—which may then pass the microbes to caretakers and, more broadly, to people who prepare and consume undercooked meat.

In agriculture, antibiotics are applied as aerosols to acres of fruit trees, for controlling or preventing bacterial infections. High concentrations may kill all the bacteria on the trees at the time of spraying, but lingering antibiotic residues can encourage the growth of resistant bacteria that later colonize the fruit during processing and shipping. The aerosols also hit more than the targeted trees. They can be carried considerable distances to other trees and food plants, where they are too dilute to eliminate full-blown infections but are still capable of killing off sensitive bac-

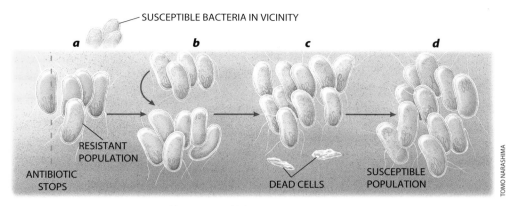

SUSCEPTIBLE BACTERIA IN VICINITY

a   b   c   d

RESISTANT POPULATION

ANTIBIOTIC STOPS

DEAD CELLS

SUSCEPTIBLE POPULATION

TOMO NARASHIMA

RESISTANT POPULATION of bacteria will disappear naturally only if susceptible bacteria live in the vicinity. After antibiotic therapy stops (*a*), resistant bacteria can persist for a while. If susceptible bacteria are nearby, however, they may recolonize the individual (*b*). In the absence of the drug, the susceptible bugs will have a slight survival advantage because they do not have to expend energy maintaining resistance genes. After a time, then, they may outcompete the resistant microbes (*c* and *d*). For this reason, protecting susceptible bacteria needs to be a public health priority.

teria and thus giving the edge to resistant versions. Here, again, resistant bacteria can make their way into people through the food chain, finding a home in the intestinal tract after the produce is eaten.

The amount of resistant bacteria people acquire from food apparently is not trivial. Denis E. Corpet of the National Institute for Agricultural Research in Toulouse, France, showed that when human volunteers went on a diet consisting only of bacteria-free foods, the number of resistant bacteria in their feces decreased 1,000-fold. This finding suggests that we deliver a supply of resistant strains to our intestinal tract whenever we eat raw or undercooked items. These bacteria usually are not harmful, but they could be if by chance a disease-causing type contaminated the food.

The extensive worldwide exploitation of antibiotics in medicine, animal care and agriculture constantly selects for strains of bacteria that are resistant to the drugs. Must all antibiotic use be halted to stem the rise of intractable bacteria? Certainly not. But if the drugs are to retain their power over pathogens, they have to be used more responsibly. Society can accept some increase in the

fraction of resistant bacteria when a disease needs to be treated; the rise is unacceptable when antibiotic use is not essential.

Reversing Resistance

A number of corrective measures can be taken right now. As a start, farmers should be helped to find inexpensive alternatives for encouraging animal growth and protecting fruit trees. Improved hygiene, for instance, could go a long way to enhancing livestock development.

The public can wash raw fruit and vegetables thoroughly to clear off both resistant bacteria and possible antibiotic residues. When they receive prescriptions for antibiotics, they should complete the full course of therapy (to ensure that all the pathogenic bacteria die) and should not "save" any pills for later use. Consumers also should refrain from demanding antibiotics for colds and other viral infections and might consider seeking nonantibiotic therapies for minor conditions, such as certain cases of acne. They can continue to put antibiotic ointments on small cuts, but they should think twice about routinely us-

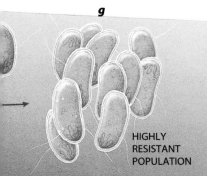

g

HIGHLY RESISTANT POPULATION

ANTIBIOTIC USE SELECTS—promotes the evolution and growth of—bacteria that are insensitive to that drug. When bacteria are exposed to an antibiotic (*a*), bacterial cells that are susceptible to the drug will die (*b*), but those with some insensitivity may survive and grow (*c*) if the amount of drug delivered is too low to eliminate every last cell. As treatment continues, some of the survivors are likely to acquire even stronger resistance (*d*)—either through a genetic mutation that generates a new resistance trait or through gene exchange with newly arriving bacteria. These resistant cells will then evade the drug most successfully (*e*) and will come to predominate (*f* and *g*).

SCIENTIFIC AMERICAN   Updated July 2007   67

ing hand lotions and a proliferation of other products now imbued with antibacterial agents. New laboratory findings indicate that certain of the bacteria-fighting chemicals being incorporated into consumer products can select for bacteria resistant both to the antibacterial preparations and to antibiotic drugs [see box on page 64].

Physicians, for their part, can take some immediate steps to minimize any resistance ensuing from required uses of antibiotics. When possible, they should try to identify the causative pathogen before beginning therapy, so they can prescribe an antibiotic targeted specifically to that microbe instead of having to choose a broad-spectrum product. Washing hands after seeing each patient is a major and obvious, but too often overlooked, precaution.

To avoid spreading multidrug-resistant infections between hospitalized patients, hospitals place the affected patients in separate rooms, where they are seen by gloved and gowned health workers and visitors. This practice should continue.

Having new antibiotics could provide more options for treatment. In the 1980s pharmaceutical manufacturers, thinking infectious diseases were essentially conquered, cut back severely on searching for additional antibiotics. At the time, if one drug failed, another in the arsenal would usually work (at least in the industrial nations, where supplies are plentiful). Now that this happy state of affairs is coming to an end, researchers are searching for novel antibiotics again. Regrettably, though, few drugs are likely to pass soon all technical and regulatory hurdles needed to reach the market. Furthermore, those that are close to being ready are structurally similar to existing antibiotics; they could easily encounter bacteria that already have defenses against them.

With such concerns in mind, scientists are also working on strategies that will give new life to existing antibiotics. Many bacteria evade penicillin and its relatives by switching on an enzyme, penicillinase, that degrades those compounds. An antidote already on pharmacy shelves contains an inhibitor of penicillinase; it prevents the breakdown of penicillin and so frees the antibiotic to work normally. In one of the strategies under study, my laboratory at Tufts University is developing a compound to jam a microbial pump that ejects tetracycline from bacteria; with the pump inactivated, tetracycline can penetrate bacterial cells effectively.

Considering the Environmental Impact

As exciting as the pharmaceutical research is, overall reversal of the bacterial resistance problem will require public health officials, physicians, farmers and others to think about the effects of antibiotics in new ways. Each time an antibiotic is delivered, the fraction of resistant bacteria in the treated individual and, potentially, in others, increases. These resistant strains endure for some time—often for weeks—after the drug is removed.

The main way resistant strains disappear is by squaring off with susceptible versions that persist in—or enter—a treated person after antibiotic use has stopped. In the absence of antibiotics, susceptible strains have a slight survival advantage, because the resistant bacteria have to divert some of their valuable energy from reproduction to maintaining antibiotic-fighting traits. Ultimately, the susceptible microbes will win out, if they are available in the first place and are not hit by more of the drug before they can prevail.

Correcting a resistance problem, then, requires both improved management of antibiotic use and restoration of the environmental bacteria susceptible to these drugs. If all reservoirs of susceptible bacteria were eliminated, resistant forms would face no competition for survival and would persist indefinitely.

In the ideal world, public health officials would know the extent of antibiotic resistance in both the infectious and benign bacteria in a community. To treat a specific pathogen, physicians would favor an antibiotic most likely to encounter little resistance from any bacteria in the community. And they would deliver enough antibiotic to clear the in-

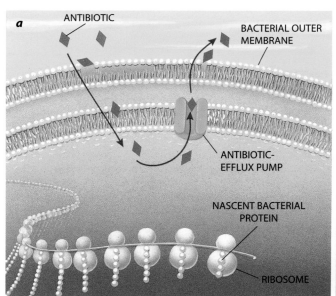
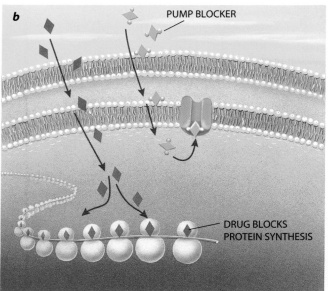

ONE PHARMACEUTICAL STRATEGY for overcoming resistance capitalizes on the discovery that some bacteria defeat certain antibiotics, such as tetracycline, by pumping out the drugs (a). To combat that ploy, investigators are devising compounds that would jam the pumps (b), thereby freeing the antibiotics to function effectively. In the case of tetracycline, the antibiotic works by interfering with the ribosomes that manufacture bacterial proteins.

Some Actions Physicians and Consumers Can Take to Limit Resistance

The easy accessibility to antibiotics parodied in the cartoon is a big contributor to antibiotic resistance. This list suggests some immediate steps that can help control the problem. —*S.B.L.*

Physicians
- Wash hands thoroughly between patient visits.
- Do not accede to patients' demands for unneeded antibiotics.
- When possible, prescribe antibiotics that target only a narrow range of bacteria.
- Isolate hospital patients with multidrug-resistant infections.
- Familiarize yourself with local data on antibiotic resistance.

Consumers
- Do not demand antibiotics.
- When given antibiotics, take them exactly as prescribed and complete the full course of treatment; do not hoard pills for later use.
- Wash fruits and vegetables thoroughly; avoid raw eggs and undercooked meat, especially in ground form.
- Use soaps and other products with antibacterial chemicals only when protecting a sick person whose defenses are weakened.

MICK STEVENS © 1998 FROM THE NEW YORKER COLLECTION, ALL RIGHTS RESERVED

"Don't forget to take a handful of our complimentary antibiotics on your way out."

---

fection completely but would not prolong therapy so much as to destroy all susceptible bystanders in the body.

Prescribers would also take into account the number of other individuals in the setting who are being treated with the same antibiotic. If many patients in a hospital ward were being given a particular antibiotic, this high density of use would strongly select for bacterial strains unsubmissive to that drug and would eliminate susceptible strains. The ecological effect on the ward would be broader than if the total amount of the antibiotic were divided among just a few people. If physicians considered the effects beyond their individual patients, they might decide to prescribe different antibiotics for different patients, or in different wards, thereby minimizing the selective force for resistance to a single medication.

Put another way, prescribers and public health officials might envision an "antibiotic threshold": a level of antibiotic usage able to correct the infections within a hospital or community but still falling below a threshold level that would

strongly encourage propagation of resistant strains or would eliminate large numbers of competing, susceptible microbes. Keeping treatment levels below the threshold would ensure that the original microbial flora in a person or a community could be restored rapidly by susceptible bacteria in the vicinity after treatment ceased.

The problem, of course, is that no one yet knows how to determine where that threshold lies, and most hospitals and communities lack detailed data on the nature of their microbial populations. Yet with some dedicated work, researchers should be able to obtain both kinds of information.

Control of antibiotic resistance on a wider, international scale will require cooperation among countries around the globe and concerted efforts to educate the world's populations about drug resistance and the impact of improper antibiotic use. As a step in this direction, various groups are now attempting to track the emergence of resistant bacterial strains. For example, an international organization, the Alliance for the Pru-

dent Use of Antibiotics (75 Kneeland St., Boston, MA 02111), has been monitoring the worldwide emergence of such strains since 1981. The group shares information with members in more than 90 countries. It also produces educational brochures for the public and for health professionals.

The time has come for global society to accept bacteria as normal, generally beneficial components of the world and not try to eliminate them—except when they give rise to disease. Reversal of resistance requires a new awareness of the broad consequences of antibiotic use— a perspective that concerns itself not only with curing bacterial disease at the moment but also with preserving microbial communities in the long run, so that bacteria susceptible to antibiotics will always be there to outcompete resistant strains. Similar enlightenment should influence the use of drugs to combat parasites, fungi and viruses. Now that consumption of those medicines has begun to rise dramatically, troubling resistance to these other microorganisms has begun to climb as well. ◼

The Author

STUART B. LEVY is professor of molecular biology and microbiology, professor of medicine and director of the Center for Adaptation Genetics and Drug Resistance at the Tufts University School of Medicine. He is also president of the Alliance for the Prudent Use of Antibiotics and past president of the American Society for Microbiology.

Further Reading

THE ANTIBIOTIC PARADOX: HOW MIRACLE DRUGS ARE DESTROYING THE MIRACLE. S. B. Levy. Plenum Publishers, 1992.
DRUG RESISTANCE: THE NEW APOCALYPSE. Special issue of *Trends in Microbiology*, Vol. 2, No. 10, pages 341–425; October 1, 1994.
ANTIBIOTIC RESISTANCE: ORIGINS, EVOLUTION, SELECTION AND SPREAD. Edited by D. J. Chadwick and J. Goode. John Wiley & Sons, 1997.

Questions for Review

"The Challenge of Antibiotic Resistance"

By Stuart B. Levy

TESTING YOUR COMPREHENSION

1. All of the following are problems associated with antibiotic resistance except which one?
   a. The prevalence of resistance genes
   b. The extent of antibiotic use
   c. Antibiotic-resistant pathogens being more virulent than susceptible bacteria
   d. None of the above, all are true

2. The term antibiotic includes all of the following except which one?
   a. Chemicals used to treat any disease
   b. Completely synthetic antimicrobials
   c. Natural antimicrobial substances that have been chemically altered
   d. Natural substances made by living organisms that inhibit growth of bacteria

3. Which one of the following is not an antibiotic-resistance mechanism employed by bacteria?
   a. Altered antibiotic-binding sites
   b. Faster growth rate
   c. Enzymes that degrade antibiotics
   d. Export pumps that push antibiotics out of cells

4. Bacteria acquire antibiotic resistance by
   a. conjugation.
   b. transduction.
   c. transformation.
   d. all of the above.

5. Plasmid-mediated antibiotic resistance is most likely passed between bacteria by
   a. conjugation.
   b. transduction.
   c. transformation.
   d. all of the above.

6. Penicillin was intensively used in an effort to prevent bacterial pneumonia. The number of pneumonia deaths immediately decreased. Each year thereafter the mortality rate from pneumonia gradually increased until ten years later it was as high as when the use of penicillin began. Which one of the following is the most likely explanation?
   a. Bacteria from other areas moved in and replaced those killed by penicillin.
   b. The few bacteria that were affected by penicillin but survived developed resistance to penicillin that they passed on to their descendants.
   c. The penicillin killed susceptible bacteria but the few that were naturally resistant lived and reproduced.
   d. The penicillin caused new mutations to occur in the surviving bacteria and this resulted in resistance to penicillin.

7. The public can help reduce antibiotic resistance by
   a. taking antibiotics as prescribed.
   b. using antibacterial household products.
   c. using antibiotic ointments.
   d. all of the above.

8. Which one of the following contributes to antibiotic resistance?
   a. Stopping your 10-day prescription because you feel better after three days of treatment
   b. Taking an antibiotic for a cold
   c. Using a friend's leftover antibiotic
   d. All of the above

9. New resistance genes arise from
   a. antibiotics in cattle feed.
   b. conjugation.
   c. mutations.
   d. transposons.

10. Chlorine bleach (NaClO) and ammonia (NH_3) are better household disinfectants than triclosan and quaternary ammonium compounds because NaClO and NH_3
    a. readily evaporate.
    b. don't kill bacteria.
    c. don't smell as bad.
    d. kill bacteria.

MICROBIOLOGY IN SOCIETY

1. Antibiotics are not present in meat or milk. Why is antibiotic use in animal feed contributing to antibiotic resistance?

2. You have flulike symptoms and your friend offers you her leftover penicillin tablets. You are not allergic to penicillin but decline the offer. How will you explain this to your friend when she asks why?

3. Your local municipal water district has proposed a ban on household triclosan-containing products. Explain the rationale for such a ban.

THINKING ABOUT MICROBIOLOGY

1. Avoparcin, a glycopeptide antibiotic, was widely used in animal feed in Denmark. In 1995, the percent of vancomycin-resistant *Enterococcus faecalis* (VRE) isolated from pigs was 100. Vancomycin is another glycopeptide. The use of avoparcin was banned in Denmark in 1995. In 2002, Danish scientists reported 20% of their *E. faecalis* isolates was vancomycin resistant. Explain why the percent of VRE decreased. Why is VRE in pigs important to humans?

2. Explain what the author meant by "the entire bacterial world can be thought of as one huge multicellular organism."

3. What role do transposons play in antibiotic resistance?

WRITING ABOUT MICROBIOLOGY

1. Use antibiotic resistance to explain natural selection.

2. Describe the methods by which bacteria acquire antibiotic resistance from other bacteria.

3. *Acinetobacter* spp. are unlikely pathogens. They are found in soil and water and have an optimum growth temperature of 33-35°C. In the last ten years, *Acinetobacter* spp. have been implicated in a variety of nosocomial infections, especially pneumonia in patients confined to hospital intensive care units. In one hospital, an outbreak of imipenem-resistant *Acinetobacter* infections followed increased use of imipenem to treat an outbreak of cephalosporin-resistant *Klebsiella pneumoniae*. Explain the emergence of *Acinetobacter* as a pathogen. How can *Acinetobacter* infections be prevented?

Answers can be found on The Microbiology Place website. Go to www.microbiologyplace.com, click on the cover of your textbook, and type in your login name and password (using the access code found in the front pages of your textbook). Then, click on Current Issues Magazine Answers on the left navigation bar.

Emerging Viruses

Hemorrhagic fever viruses are among the most dangerous biological agents known. New ones are discovered every year, and artificial as well as natural environmental changes are favoring their spread

by Bernard Le Guenno

Data in this article have been updated by Christine L. Case, PhD, at Skyline College to include information occurring after the original 1995 publication date. The author, Bernard Le Guenno, was not available for review or comment on these updates.

In May 1993 a young couple in New Mexico died just a few days apart from acute respiratory distress. Both had suddenly developed a high fever, muscular cramps, headaches and a violent cough. Researchers promptly started looking into whether similar cases had been recorded elsewhere. Soon 24 were identified, occurring between December 1, 1992, and June 7, 1993, in New Mexico, Colorado and Nevada. Eleven of these patients had died.

Bacteriological, parasitological and virological tests conducted in the affected states were all negative. Samples were then sent to the Centers for Disease Control and Prevention (CDC) in Atlanta. Tests for all known viruses were conducted, and researchers eventually detected in the serum of several patients antibodies against a class known as hantaviruses. Studies using the techniques of molecular biology showed that the patients had been infected with a previously unknown type of hantavirus, now called Sin Nombre (Spanish for "no name").

New and more effective analytical techniques are identifying a growing number of infective agents. Most are viruses that 20 years ago would probably have passed unnoticed or been mistaken for other, known types. The Sin Nombre infections were not a unique occurrence. In 1994, a researcher at the Yale University School of Medicine was accidentally infected with Sabià, a virus first isolated in 1990 from an agricultural engineer who died from a sudden illness in the state of São Paulo, Brazil.

Sabià and Sin Nombre both cause illnesses classified as hemorrhagic fevers. Patients initially develop a fever, followed by a general deterioration in health during which bleeding often occurs. Superficial bleeding reveals itself through skin signs, such as petechiae (tiny releases of blood from vessels under the skin surface), bruises or purpura (characteristic purplish discolorations). Other cardiovascular, digestive, renal and neurological complications can follow. In the most serious cases, the patient dies of massive hemorrhag-

es or sometimes multiple organ failure.

Hemorrhagic fever viruses are divided into several families. The flaviviruses have been known for the longest. They include the Amaril virus that causes yellow fever and is transmitted by mosquitoes, as well as other viruses responsible for mosquito- and tick-borne diseases, such as dengue. Viruses that have come to light more recently belong to three other families: arenaviruses, bunyaviruses (a group that includes the hantaviruses) and filoviruses. They have names like Puumala, Guanarito and Ebola, taken from places where they first caused recognized outbreaks of disease.

All the arenaviruses and the bunyaviruses responsible for hemorrhagic fevers circulate naturally in various populations of animals. It is actually uncommon for them to spread directly from person to person. Epidemics are, rather, linked to the presence of animals that serve as reservoirs for the virus and sometimes as vectors that help to transfer it to people. Various species of rodent are excellent homes for these viruses, because the rodents show no signs when infected. Nevertheless, they shed viral particles throughout their lives in feces and, particularly, in urine. The filoviruses, for their part, are still a mystery: we do not know how they are transmitted.

Hemorrhagic fever viruses are among the most threatening examples of what are commonly termed emerging pathogens. They are not really new. Mutations or genetic recombinations between existing viruses can increase virulence, but what appear to be novel viruses are generally viruses that have existed for millions of years and merely come to light when environmental conditions change. The changes allow the virus to multiply and spread in host organisms. New illnesses may then sometimes become apparent.

Improvements in Diagnosis

The seeming emergence of new viruses is also helped along by rapid advances in the techniques for virological identification. The first person diagnosed with Sabià in São Paulo (called the index case) was originally thought to be suffering from yellow fever. The agent actually responsible was identi-

fied only because a sample was sent to a laboratory equipped for the isolation of viruses. That rarely happens, because most hemorrhagic fever viruses circulate in tropical regions, where hospitals generally have inadequate diagnostic equipment and where many sick people are not hospitalized. Even so, the rapid identification of Sin Nombre was possible only because of several years of work previously accumulated on hantaviruses.

Hantaviruses typically cause an illness known as hemorrhagic fever with renal syndrome; it was described in a Chinese medical text 1,000 years ago. The West first became interested in this illness during the Korean War, when more than 2,000 United Nations troops suffered from it between 1951 and 1953. Despite the efforts of virologists, it was not until 1976 that the agent was identified in the lungs of its principal reservoir in Korea, a field mouse. It took more than four years to isolate the virus, to adapt it to a cell culture and to prepare a reagent that permitted a diagnostic serological test, essential steps in the study of a virus. It was named Hantaan, for a river in Korea. The virus also circulates in Japan and Russia, and a similar virus that produces an illness just as serious is found in the Balkans.

A nonfatal form exists in Europe. It was described in Sweden in 1934 as the "nephritic epidemic," but its agent was not identified until 1980, when it was detected in the lungs of the bank vole. Isolated in 1983 in Finland, the virus was named Puumala for a lake in that country. Outbreaks occur regularly in northwestern Europe. Since 1977, 505 cases have been recorded in northeastern France alone. The number of cases seems to be increasing, but this is probably because doctors are using more biological tests than formerly, and because the tests in recent years have become more sensitive.

Thus, it is only for about a decade that we have had the reagents necessary to identify hantaviruses. Thanks to these reagents and a research technique that spots antibodies marking recent infections, scientists at the CDC in 1993 were quickly on the track of the disease. The presence of specific antibodies is not always definite proof of an infection by the corresponding pathogen, however. False positive reactions and cross-reactions caused by the presence of antibodies shared by different viruses are possible. A more recent technology, based on the polymerase chain reaction, permits fragments of genes to be amplified (or duplicated) and sequenced. It provided confirmation that

A. B. DOWSETT SPL/Photo Researchers, Inc.

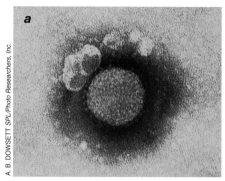

SCOTT CAMAZINE Photo Researchers, Inc.

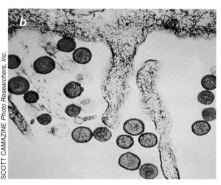

A. B. DOWSETT SPL/Photo Researchers, Inc.

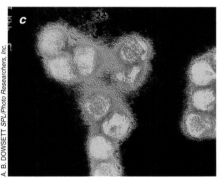

SCIENCE SOURCE/PHOTO RESEARCHERS, INC.

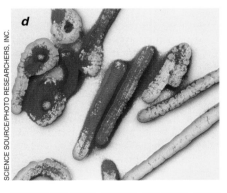

HEMORRHAGIC FEVER VIRUSES vary greatly in appearance under the electron microscope. Lassa (*a*), found in Africa, is an arenavirus, a kind that is typically spherical. Hantaviruses (*b*) cause diseases of different varieties in many regions of the world. Tick-borne encephalitis virus (*c*) is an example of a flavivirus, a group that includes yellow fever and dengue. Ebola (*d*) is one of the filoviruses, so called because of their filamentous appearance. The images have been color-enhanced.

CONGOLESE RED CROSS members bury victims of the Ebola virus in Kikwit in 1995. At least 250 died in the epidemic. Poor medical hygiene and unsafe funeral practices helped to propagate the infection.

the patients were indeed infected with hantaviruses. The identification of Sin Nombre took no more than eight days.

The Infective Agents

The primary cause of most outbreaks of hemorrhagic fever viruses is ecological disruption resulting from human activities. The expansion of the world population perturbs ecosystems that were stable a few decades ago and facilitates contacts with animals carrying viruses pathogenic to humans. This was true of the arenavirus Guanarito, discovered in 1989 in an epidemic in Venezuela. The first 15 cases were found in a rural community that had started to clear a forested region in the center of the country. The animal reservoir is a species of cotton rat; workers had stirred up dust that had been contaminated with dried rat urine or excrement—one of the most frequent modes of transmission. Subsequently, more than 100 additional cases were diagnosed in the same area.

Other arenaviruses responsible for hemorrhagic fevers have been known for a long time—for example, Machupo, which appeared in Bolivia in 1952, and Junín, identified in Argentina in 1958. Both those viruses can reside in species of rodents called vesper mice; the Bolivian species enters human dwellings. Until recently, an extermination campaign against the animals had prevented any human infections with Machupo since 1974. After a lull of 20 years, however, this virus reappeared, in the same place: seven people, all from one family, were infected during the summer of 1994.

Junín causes Argentinian hemorrhagic fever, which appeared at the end of the 1940s in the pampas west of Buenos Aires. The cultivation of large areas of maize supported huge populations of the species of vesper mice that carry this virus and multiplied contacts between these rodents and agricultural workers. Today mechanization has put the operators of agricultural machinery on the front line: combine harvesters not only suspend clouds of infective dust, they also create an aerosol of infective blood when they accidentally crush the animals.

The arenavirus Sabià has, so far as is known, claimed only one life, but other cases have in all probability occurred in Brazil without being diagnosed. There is a real risk of an epidemic if agricultural practices bring the inhabitants of São Paulo into contact with rodent vectors. In Europe, the main reservoirs of the hantavirus Puumala—the bank vole and yellow-necked field mouse—are

Global Reach of Hemorrhagic Fever Viruses

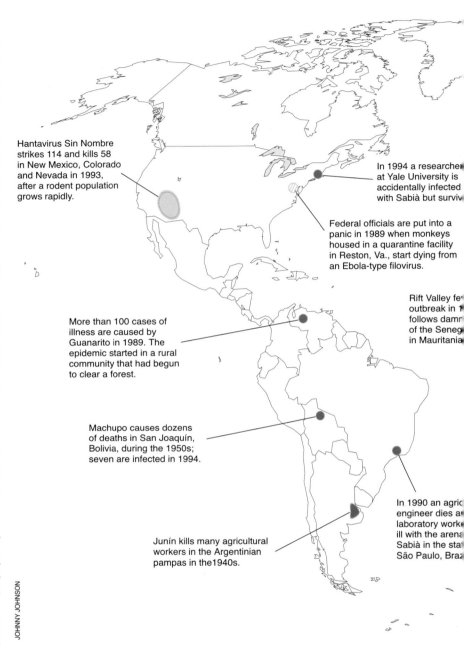

Hantavirus Sin Nombre strikes 114 and kills 58 in New Mexico, Colorado and Nevada in 1993, after a rodent population grows rapidly.

In 1994 a researcher at Yale University is accidentally infected with Sabià but surviv[...]

Federal officials are put into a panic in 1989 when monkeys housed in a quarantine facility in Reston, Va., start dying from an Ebola-type filovirus.

Rift Valley fe[...] outbreak in 1[...] follows damm[...] of the Seneg[...] in Mauritania[...]

More than 100 cases of illness are caused by Guanarito in 1989. The epidemic started in a rural community that had begun to clear a forest.

Machupo causes dozens of deaths in San Joaquín, Bolivia, during the 1950s; seven are infected in 1994.

In 1990 an agric[...] engineer dies a[...] laboratory work[...] ill with the aren[...] Sabià in the sta[...] São Paulo, Braz[...]

Junín kills many agricultural workers in the Argentinian pampas in the1940s.

JOHNNY JOHNSON

woodland animals. The most frequent route of contamination there is inhalation of contaminated dust while handling wood gathered in the forest or while working in sheds and barns.

Humans are not always the cause of dangerous environmental changes. The emergence of Sin Nombre in the U.S. resulted from heavier than usual rain and snow during spring 1993 in the mountains and deserts of New Mexico, Nevada and Colorado. The principal animal host of Sin Nombre is the deer mouse, which lives on pine kernels: the excep-

tional humidity favored a particularly abundant crop, and so the mice proliferated. The density of the animals multiplied 10-fold between 1992 and 1993.

Transmission by Mosquitoes

Some bunyaviruses are carried by mosquitoes rather than by rodents. Consequently, ecological perturbations such as the building of dams and the expansion of irrigation can encourage these agents. Dams raise the water table, which favors the multiplication of

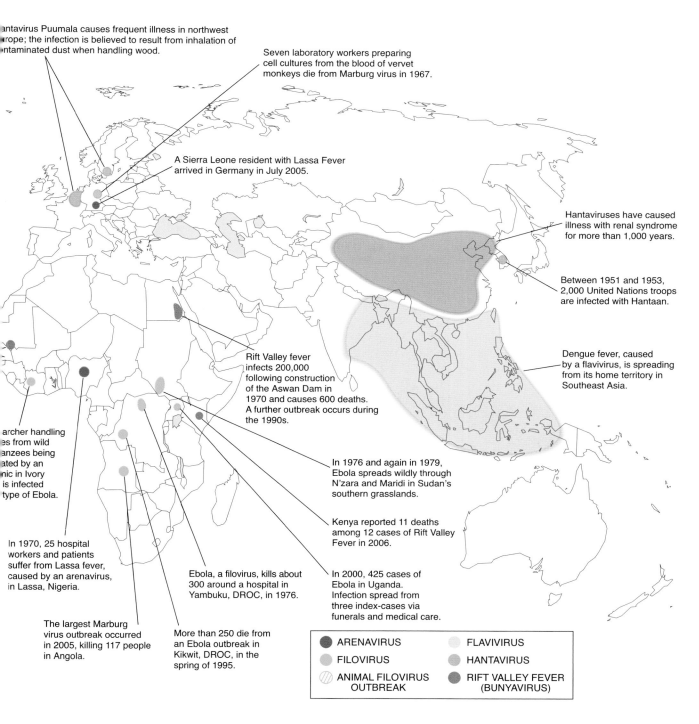

Hantavirus Puumala causes frequent illness in northwest Europe; the infection is believed to result from inhalation of contaminated dust when handling wood.

Seven laboratory workers preparing cell cultures from the blood of vervet monkeys die from Marburg virus in 1967.

A Sierra Leone resident with Lassa Fever arrived in Germany in July 2005.

Hantaviruses have caused illness with renal syndrome for more than 1,000 years.

Between 1951 and 1953, 2,000 United Nations troops are infected with Hantaan.

Rift Valley fever infects 200,000 following construction of the Aswan Dam in 1970 and causes 600 deaths. A further outbreak occurs during the 1990s.

Dengue fever, caused by a flavivirus, is spreading from its home territory in Southeast Asia.

...archer handling ...es from wild ...anzees being ...ated by an ...nic in Ivory ...is infected ...type of Ebola.

In 1976 and again in 1979, Ebola spreads wildly through N'zara and Maridi in Sudan's southern grasslands.

Kenya reported 11 deaths among 12 cases of Rift Valley Fever in 2006.

In 1970, 25 hospital workers and patients suffer from Lassa fever, caused by an arenavirus, in Lassa, Nigeria.

Ebola, a filovirus, kills about 300 around a hospital in Yambuku, DROC, in 1976.

In 2000, 425 cases of Ebola in Uganda. Infection spread from three index-cases via funerals and medical care.

The largest Marburg virus outbreak occurred in 2005, killing 117 people in Angola.

More than 250 die from an Ebola outbreak in Kikwit, DROC, in the spring of 1995.

- ● ARENAVIRUS
- ● FILOVIRUS
- ⦸ ANIMAL FILOVIRUS OUTBREAK
- ○ FLAVIVIRUS
- ● HANTAVIRUS
- ● RIFT VALLEY FEVER (BUNYAVIRUS)

the insects and also brings humans and animals together in new population centers. These two factors probably explain two epidemics of Rift Valley fever in Africa: one in 1977 in Egypt and the other in 1987 in Mauritania.

The virus responsible was recognized as long ago as 1931 as the cause of several epizootics, or animal epidemics, among sheep in western and South Africa. Some breeders in contact with sick or dead animals became infected, but at the time the infection was not serious in humans. The situation became more

grim in 1970. After the construction of the Aswan Dam, there were major losses of cattle; of the 200,000 people infected, 600 died. In 1987 a minor epidemic followed the damming of the Senegal River in Mauritania.

Rift Valley fever virus is found in several species of mosquitoes, notably those of the genus *Aedes*. The females transmit the virus to their eggs. Under dry conditions the mosquitoes' numbers are limited, but abundant rain or irrigation allows them to multiply rapidly. In the course of feeding on blood,

they then transmit the virus to humans, with cattle acting as incubators.

Contamination by Accident

Although important, ecological disturbances are not the only causes of the emergence of novel viruses. Poor medical hygiene can foster epidemics. In January 1969 in Lassa, Nigeria, a nun who worked as a nurse fell ill at work. She infected, before dying, two other nuns, one of whom died. A year later an epidemic broke out in the same hospital. An inquiry found that 17 of the 25 persons infected had probably been in the room where the first victim had been hospitalized. Lassa is classed as an arenavirus.

Biological industries also present risks. Many vaccines are prepared from animal cells. If the cells are contaminated, there is a danger that an unidentified virus may be transmitted to those vaccinated. It was in this way that in 1967 a culture of contaminated blood cells allowed the discovery of a new hemorrhagic fever and a new family of viruses, the filoviruses.

The place was Marburg, Germany, where 25 people fell ill after preparing cell cultures from the blood of vervet monkeys. Seven died. Other cases were reported simultaneously in Frankfurt and in Yugoslavia, all in laboratories that had received monkeys from Uganda. The monkeys themselves also died, suggesting that they are not the natural reservoir of Marburg virus. Two outbreaks

five of them had received an injection in this hospital. The epidemic led to the identification of a new virus, Ebola.

The Marburg and Ebola viruses are classified as filoviruses, so called because under the electron microscope they can be seen as filamentous structures as much as 1,500 nanometers in length (the spherical particle of an arenavirus, for comparison, is about 300 nanometers in diameter). These two representatives of the filovirus family are exceedingly dangerous. In 1989 specialists at the CDC were put in a panic when they learned that crab-eating macaques from the Philippines housed in an animal quarantine facility in Reston, Va., were dying from an infection caused by an Ebola-type filovirus. The

AGRICULTURAL WORKERS in some parts of the world are at risk of infection by arenaviruses, which are often carried by rodents. Machinery stirs up dried rodent urine containing the viruses and can create an aerosol of infective blood if the animals are accidentally crushed.

of natural infection with Marburg have been reported in Africa, but neither the reservoir nor the natural modes of transmission have been discovered. What is clear is that Marburg can propagate in hospitals: secondary cases have occurred among medical personnel.

In 1976 two epidemics of fever caused by a different virus occurred two months apart in the south of Sudan and in northern Democratic Republic of Congo (DROC). In DROC, around Yambuku Hospital, by the Ebola River, 318 cases were counted, and 280 persons died. Eighty-

virus was also isolated from other animal facilities that had received monkeys from the Philippines. No human illnesses were recorded in the wake of this epizootic, however, which demonstrates that even closely related viruses can vary widely in their effects.

In January 1995 we isolated a previously unknown type of Ebola from a patient who had infected herself handling samples from wild chimpanzees that were being decimated by a strange epidemic. That the chimpanzees, from Ivory Coast, succumbed is further evidence that primates are not filoviruses' natural reservoir, which has not yet been identified. Although Marburg has infected few people, Ebola surfaced

again to cause a human epidemic in the DROC in 1995 [see box, pp 78 - 79] and in Uganda in 2000.

A Shifting, Hazy Target

The extreme variability and speed of evolution found among hemorrhagic fever viruses are rooted in the nature of their genetic material. Hemorrhagic fever viruses, like many other types, generally have genes consisting of ribonucleic acid, or RNA, rather than the DNA employed by most living things. The RNA of these viruses is "negative stranded"—before it can be used to make viral proteins in an infected cell, it must be converted into a positive

BARRY ROSS

RIFT VALLEY FEVER VIRUS, a bunyavirus, is transmitted by mosquitoes from cattle and sheep to humans. Dams allow proliferation of the insects by raising the water table and bring people and herds together in new locations, causing epidemics.

and by an enzyme called RNA polymerase. RNA polymerases cause fairly frequent errors during this process. Because the errors are not corrected, an infected cell gives rise to a heterogeneous population of viruses resulting from the accumulating mutations. The existence of such "quasispecies" explains the rapid adaptation of these viruses to environmental changes. Some adapt to invertebrates and others to vertebrates, and they confound the immune systems of their hosts. Pathogenic variants can easily arise.

There is another source of heterogeneity, too. A characteristic common to arenaviruses and bunyaviruses is that they have segmented genomes. (The bunyaviruses have three segments of RNA, arenaviruses two.) When a cell is infected by two viruses of the same general class, they can then recombine so that segments from one become linked to segments from the other, giving rise to new viral types called reassortants.

Although we have a basic appreciation of the composition of these entities, we have only a poor understanding of how they cause disease. Far beyond the limited means of investigation in local tropical hospitals, many of these viruses are so hazardous they cannot be handled except in laboratories that conform to very strict safety requirements. There are only a few such facilities in the world, and not all of them have the required equipment. Although it is relatively straightforward to handle the agents safely in culture flasks, it is far more dangerous to handle infected monkeys: researchers risk infection from being scratched or bitten by sick animals. Yet the viruses cannot be studied in more common laboratory animals such as rats, because these creatures do not become ill when infected.

We do know that hemorrhagic fever viruses have characteristic effects on the body. They cause a diminution in the number of platelets, the principal cells of the blood-clotting system. But this diminution, called thrombocytopenia, is not sufficient to explain the hemorrhagic symptoms. Some hemorrhagic fever viruses destroy infected cells directly; others perturb the immune system and affect cells' functioning.

Among the first group, the cytolytic viruses, are the bunyaviruses that cause a disease called Crimean-Congo fever and Rift Valley fever; the filoviruses Marburg and Ebola; and the prototype of hemorrhagic fever viruses, the flavivirus Amaril. Their period of incubation is generally short, often less than a week. Serious cases are the result of an attack on several organs, notably the liver. When a large proportion of liver cells are destroyed, the body cannot produce enough coagulation factors, which partly explains the hemorrhagic symptoms. The viruses also modify the inner surfaces of blood vessels in such a way that platelets stick to them. This clotting inside vessels consumes additional coagulation factors. Moreover, the cells lining the vessels are forced apart, which can lead to the escape of plasma or to uncontrolled bleeding, causing edema, an accumulation of fluid in the tissue, or severely lowered blood pressure.

The arenaviruses fall into the noncytolytic group. Their period of incubation is longer, and although they invade most of the tissues in the body, they do not usually cause gross lesions. Rather the viruses inhibit the immune system, which delays the production of antibodies until perhaps a month after the first clinical signs of infection. Arenaviruses

Ebola's Unanswered Questions

by Laurie Garrett

In 1995 in Kikwit, DROC, Ebola proved once again that despite the agonizing and usually fatal illness it provokes, the microbe cannot in its present incarnation spread far—unless humans help it to do so. The virus is too swiftly lethal to propagate by itself. In the early waves of an epidemic, it kills more than 92 percent of those it infects, usually within a couple of weeks. Such rapidity affords the microbe little opportunity to spread unaided, given the severity of the illness that it causes.

In each of the four known Ebola epidemics during the past 30 years, people have helped launch the virus from its obscure rain forest or savanna host into human populations. In 1976 in Yambuku, an area of villages in DROC's northern rain forest, the virus's appearance was multiplied dozens of times over by Belgian nuns at a missionary clinic who repeatedly used unsterilized syringes in some 300 patients every day. One day someone arrived suffering from the then unknown Ebola fever and was treated with injections for malaria. The syringes efficiently amplified the viral threat.

In both 1976 and 1979, humans helped the virus spread wildly in N'zara and Maridi, in the Sudan's remote southern grasslands. Improper hospital hygiene again played a key role, and local burial practices, which required the manual removal of viscera from cadavers, compounded the disaster.

Medical and funeral settings were likewise crucial in Kikwit in early 1995. Infections spread via bodily fluids among those who tended the dying and washed and dressed the cadavers. The major amplification event that seems to have started the epidemic, early in the new year, was an open casket funeral. The deceased, Gaspard Menga, probably acquired his infection gathering firewood in a nearby rain forest. The virus spread rapidly to 13 members of the Menga family who had cared for the ailing man or touc[hed him] in farewell, a common practice in the region[. Among] those who got Ebola from Menga.

A second amplification event occurred in Ma[ridi? Kik]wit General Hospital. Overrun by cases of inc[reasing] diarrhea, hospital officials thought they were [facing a] strain of bacteria. The doctors ordered a lab[techni]cian to draw blood samples from patients and [test it] for drug resistance.

When he took ill, the hospital staff though[t the enor]mously distended stomach and high fever wer[e signs of] typhus infection and performed surgery to sta[bilize him]. The first procedure was an appendectomy. Th[en came] horror. When the physicians and nurses ope[ned the techni]cian's abdomen again for what they expecte[d to be routine] work, they were immediately drenched in bl[ood. A col]league died on the operating table from unco[ntrollable bleed]ing. The contaminated surgical team becam[e a new] wave of the epidemic.

The virus's reliance on unintended help from [human sourc]es attention to the common thread that ru[ns through all] known Ebola epidemics: poverty. All the ou[tbreaks have] been associated with abysmal medical facili[ties, with] poorly paid (or, in the case of Kikwit, unpaid) [per]sonnel had to make do with a handful of syrin[ges, scarce] surgical equipment and intermittent or nonexis[tent] water and electricity.

It seems quite possible that Ebola (and other [hemorrhagic] fever viruses) might successfully exploit simila[r conditions] occurring anywhere in the world. As air trans[port be]comes more readily available and affordable, viru[ses]

---

suppress the number of platelets only slightly, but they do inactivate them. Neurological complications are common.

Hantaviruses are like arenaviruses in that they do not destroy cells directly and also have a long period of incubation, from 12 to 21 days. They target cells lining capillary walls. Hantaan and Puumala viruses invade the cells of the capillary walls in the kidney, which results in edema and an inflammatory reaction caused by the organ's failure to work properly. Sin Nombre, in contrast, invades pulmonary capillaries and caus-es death by a different means[, due] to acute edema of the lung.

Prospects for Control

Several research groups are tryi[ng to] establish international surveilla[nce] networks that will track all emerg[ing] infectious agents. The World Health O[r]ganization has established a network for tracking hemorrhagic fever viruses and other insect-borne viruses that is particularly vigilant.

Once a virus is detected, technology holds some promise for combating it. An antiviral medication, ribavirin, proved effective during an epidemic of hantavirus in China. A huge effort is under way in Argentina to develop a vaccine to protect people against Junín.

PORTABLE ISOLATOR UNITS equipped with air filters have been maintained by the U.S. Army since 1980 for evacuating personnel carrying suspected dangerous pathogens. The equipment would be used to bring patients needing specialized care to an isolation facility at Fort Detrick, Md., but has never been called on for this mission.

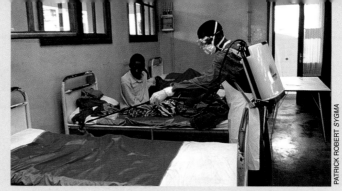

MASKED AND GLOVED health worker disinfects a bed used by a patient stricken by the Ebola virus in Kikwit, DROC.

noved around the planet. The rapid deterioration
lth and medical facilities in the former Soviet
er regions should therefore be cause for concern.
nature of the risk, of course, depends on the
biology, much of which remains mysterious.
e summer, researchers from the University of
U.S. Centers for Disease Control and Preven-
ur Institute in Paris, the National Institute of Vi-
nnesburg and the World Health Organization
t for answers to questions that have puzzled
e the first Yambuku epidemic: What are the
ints on Ebola's transmission? And where does
epidemics?

nese epidemics started among cotton factory
time scientists scoured the N'zara complex
ects or bats, but although the animals were
carried the virus. In Yambuku, suspicions fell
ain-forest animals, including monkeys. Again,
apped animals tested positive for infection.
cted during the late 1970s in conjunction with
control monkeypox found no infected pri-
nimals in central Africa.

st frequented by Gaspard Menga contained
bats, mice and snakes. Trapping efforts in the
ntually reveal Ebola's hideout. For the present,
irus's reservoir remains unknown. Also un-
her shared drinking water, foods and washing
ansmit infection.

outbreaks to date have involved transmission
control has consisted of fairly straightforward,
ts. Patients were isolated, and the citizenry in-
urn over their unwashed dead to authorities.
nts appreciated the links between tending the

sick, washing a cadaver and dying of Ebola, epidemics quickly ground to a halt.

One way that Ebola could escape such controls would be
through a major mutational event that made it more easily
transmissible. Were Ebola, or any hemorrhagic fever virus, to
acquire genetic characteristics suitable for airborne transmission, an outbreak of disease anywhere would pose a threat
to all humanity.

As far as is known, nobody has ever acquired the microbe
from inhaled droplets coughed into the air (although it can
certainly be passed in saliva during a kiss). There are usually
many genetic differences between fluid-borne microbes and
airborne ones, so it seems unlikely that the jump could be
made easily. But the question has never been specifically
studied in the case of Ebola, because research on microbes
that are found primarily in developing countries has for many
years been poorly funded.

LAURIE GARRETT is a reporter for Newsday and the author of The Coming Plague: Newly Emerging Diseases in a World Out of Balance *(Penguin USA, 1995).*

cines against the Rift Valley
imals, and against yellow fe-
ans, are already approved for
espite the existence of yellow
ine, that disease is now raging
where few are vaccinated.
approaches are constrained be-
is difficult or impossible to con-
imals that are natural reservoirs
ctors for the viruses or to predict
gical modifications that favor out-
ks of disease. There was an effec-
campaign against rodent vectors
ring the Lassa and Machupo arena-
us outbreaks, but it is not usually

possible to sustain such programs in
rural regions for long periods.

Precautions can be taken in laboratories and hospitals, which have ironically
served as amplifiers in several epidemics. In the laboratory, viruses responsible for hemorrhagic fevers must be handled in maximum confinement conditions (known in the jargon as biosafety
level 4). The laboratory must be kept at
lowered pressure, so that no potentially infectious particle can escape; the viruses themselves should be confined in
sealed systems at still lower pressure.
In hospitals, the risk of infection from a

patient is high for some viruses, so strict
safety measures must be followed: hospital personnel must wear masks, gloves
and protective clothing; wastes must be
decontaminated. A room with lowered
pressure is an additional precaution.

Since penicillin has been in widespread use, many people had started to
believe that epidemics were no longer a
threat. The global pandemic of HIV, the
virus that causes AIDS, has shown that
view to be complacent. Hemorrhagic fever viruses are indeed a cause for worry, and the avenues to reduce their toll
are still limited.

The Author

BERNARD LE GUENNO leads the national reference center for hemor-rhagic fever viruses at the Pasteur In-stitute in Paris. He graduated with a degree in pharmacology from Bor-deaux University in 1972 and has been a virologist at Pasteur since 1983. This article was adapted from one by Le Guenno in the June 1995 issue of *Pour la Science,* the French edition of *Scientific American.*

Further Reading

GENETIC IDENTIFICATION OF A HANTAVIRUS ASSOCIATED WITH AN OUTBREAK OF ACUTE RESPIRATO-RY ILLNESS. Stuart T. Nichol et al. in *Science,* Vol. 262, pages 914–917; November 5, 1993.
HANTAVIRUS EPIDEMIC IN EUROPE, 1993. B. Le Guenno, M. A. Camprasse, J. C. Guilbaut, Pascale Lanoux and Bruno Hoen in *Lancet,* Vol. 343, No. 8889, pages 114–115; January 8, 1994.
NEW ARENAVIRUS ISOLATED IN BRAZIL. Terezinha Lisieux M. Coimbra et al. in *Lancet,* Vol. 343, No. 8894, pages 391–392; February 12, 1994.
FILOVIRUSES AS EMERGING PATHOGENS. C. J. Peters et al. in *Seminars in Virology,* Vol. 5, No. 2, pages 147–154; April 1994.
ISOLATION AND PARTIAL CHARACTERISATION OF A NEW STRAIN OF EBOLA VIRUS. Bernard Le Guen-no, Pierre Formentry, Monique Wyers, Pierre Gounon, Francine Walker and Christophe Boesch in *Lancet,* Vol. 345, No. 8960, pages 1271–1274; May 20, 1995.

"Emerging Viruses"

By Bernard Le Guenno

TESTING YOUR COMPREHENSION

1. The hemorrhagic fever viruses belong to all of the following families except which one?
   a. Arenaviridae
   b. Bunyaviridae
   c. Coronaviridae
   d. Flaviviridae

2. The hemorrhagic fever viruses have
   a. single-stranded DNA.
   b. double-stranded RNA.
   c. negative single-stranded RNA.
   d. positive single-stranded RNA.

3. Emergence of hemorrhagic fever viruses occurs for all of the following reasons except which one?
   a. Cultivating crops
   b. Damming rivers
   c. Genetic engineering in labs
   d. Logging

4. The identification of several new viruses in the 1970s is primarily due to
   a. many mutations in the 1960s.
   b. new molecular biology techniques.
   c. antibody testing.
   d. the ability to go into remote regions.

5. Which of the following pairs is not correctly matched?
   a. Lassa Fever virus—Arenaviridae
   b. Rift Valley Fever virus—Bunyaviridae
   c. Marburg virus—Flaviviridae
   d. Ebola virus—Filoviridae

6. Which hemorrhagic fever virus causes acute edema of the lungs?
   a. Arenavirus Machupo
   b. Arenavirus Sabià
   c. Hantavirus Puumala
   d. Hantavirus Sin Nombre

7. Which of the following is not a[n] "amplification" factor?
   a. Contact with infected anim[als]
   b. Funeral ceremonies
   c. Improper or inadequate ma[sk] gloving
   d. Reusing syringes

8. New hemorrhagic fever viruses [arise for all] the following reasons except wh[ich one?]
   a. People go into new areas.
   b. Quasispecies exist.
   c. RNA polymerase makes mista[kes.]
   d. Segments of genomes from dif[ferent viruses] can form new viruses.

9. Which of the following provides di[rect] proof of the cause of a person's infe[ction?]
   a. Hemorrhaging
   b. PCR
   c. The presence of antibodies
   d. The presence of rodents

10. All of the following are characteristics [of] cytolytic hemorrhagic fever viruses exce[pt] which one?
    a. They cause blood to clot because platel[ets] stick to blood vessels.
    b. They cause bleeding by destroying liver ce[lls.]
    c. Incubation is less than one week.
    d. They invade capillary walls in the kidney.